Classroom Creature Culture:

Algae to Anoles

Revised Edition

A collection from the columns of

Science and Children

*by Carolyn H. Hampton,
Carol D. Hampton,
David C. Kramer
and others*

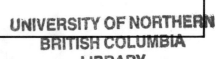

About This Collection

In the spring of 1973, I was visiting Paul Hummer's small organism culturing center in Frederick, Maryland. We talked about teacher needs and concerns for information on culturing and raising live materials in the classroom. Why did tadpoles die without maturing in an elementary classroom environment? How do you find food for field crickets? What are the best ways to collect and study pond life in the classroom? We agreed that a new column for *Science and Children* could provide the kind of "care and Maintenance" information teachers needed.

The series began in the spring of 1974, but after two articles, Paul, who was a biology teacher at Thomas Johnson High School, signed a publishing contract and had to give up writing for *Science and Children*. For the next two years, we heard from disappointed readers who longed for more of the column while we searched for a new column editor.

We found not one but two dedicated biologists willing to assume the responsibility—Carol and Carolyn Hampton, Professors of Science Education at East Carolina University in Greenville. For four years (1978-1982), this husband and wife team co-authored "Care and Maintenance" and provided our readers with excellent guidance in collecting and caring for live material. The Hamptons' series of articles closed in 1982, as they directed their talents to contract publishing demands for several textbooks and teacher workshops.

In October 1984, we launched a follow-up series, "The Classroom Animal," under the authorship of David C. Kramer, professor of biology at St. Cloud University in St. Cloud, Minnesota. The focus of Dr. Kramer's work is natural history of small animals suitable for classroom study. He also puts a special emphasis in his articles on helping teachers engender in their students an appreciation for the environment.

Classroom Creature Culture: Algae to Anoles is a collection of the best of the fine work of these authors, with some related short items by others who have contributed to *Science and Children* over the years, on the collecting and culturing of plants and animals. Although the material was intended to meet the needs of elementary school teachers, we feel confident that the techniques and content presented are appropriate at any level.

—Phyllis Marcuccio
Associate Executive Director for Publications
(former Editor of *Science and Children*)

The National Science Teachers Association
is an organization of science education professionals
and has as its purpose the stimulation, improvement,
and coordination of science teaching and learning.

Cover Photographs by Paul E. Meyers
Cover design by Brian Marquis

NSTA Stock Number PB101X
ISBN 0-87355-120-6
Printed in the United States

 Publications

Table of Contents

Living Organisms
Important Classroom Resources

This booklet is a collection of articles from *Science and Children* that in response to a growing need on the part of teachers for information about the care of specific organisms for classroom use. To aid teachers in the implementation of such elementary school science programs as Science Curriculum Improvement Study (SCIS), Elementary Science Study (ESS), and Science: A Process Approach (SAPA), techniques were researched and tried under conditions similar to those that exist in most elementary classrooms. The methods described are those that seem the most appropriate for normal classroom conditions and require the least time and resources. The organisms described in the articles are easily obtainable. Their care and maintenance are within the resources and abilities of most elementary and secondary life science teachers and students. Emphasis has been placed on plants and invertebrates that lend themselves to humane uses in the classroom setting. Even though all of the articles include most of the topics listed, Table 1 indicates which articles are particularly useful in teaching certain life science topics.

This publication is an updated version of an anthology published in 1986. Many new articles from *Science and Children* have been included, giving details about caring for a greater diversity of organisms under typical classroom conditions. Also, This new version arranges the organisms in a sequence based on evolutionary relationships, from the simpler organisms to the more complex.

Why Study Living Organisms?

Children are naturally curious about the world around them. This characteristic makes learning activities in science enjoyable and rewarding for both students and teachers. Whether they live in large cities, small towns, or rural areas, children everywhere display an interest in living things. A field trip around the school grounds or through a park or vacant lot, will provide them with many opportunities to acquire firsthand knowledge of plants and animals. Constructing and maintaining aquaria and terraria with specimens brought back from such field trips is an exciting way to illustrate the interrelationships between living things and their environment. Furthermore, a variety of well-planned investigations using easily obtainable animals and plants gives children opportunities to apply and understand the processes of the scientific method.

Many of the world's problems, which children read and hear about daily in school and on television, are the result of technology and its effect on the environment. Often, the result of this constant barrage of negativism is that students are left with a feeling of hopelessness. The study of living things both in the classroom and outdoors can develop a sensitivity, respect, and working knowledge of the requirements for all life. It can also help inculcate the attitude that humans can manipulate technology for the enhancement of their environment and the environment of other living organisms as well. The use of living materials, rather than preserved specimens, enables children to explore and devise experiments to explain the life around them. The solutions to our current problems and those in the future depend on how well teachers and schools develop a scientific literacy and ethic in today's children.

Educational Objectives

The study of living plants and animals contributes to educational objectives. Instructional activities dealing with living organisms can be integrated with many subjects and skill areas in science, mathematics, social studies, language arts, and art.

The study of live specimens contributes to the goals of education in other significant ways: It: (1) nurtures and satisfies children's natural curiosity; (2) develops a sensitivity and respect for all forms of life; (3) encourages and promotes a permanent interest in the life sciences, either as a career or leisure time activity; and (4) lays the foundation for an ecological view of life, which is basic to the solution of many environmental problems.

When living organisms are maintained in the classroom, children become aware of the conditions under which animals and plants thrive as well as those under which they will perish. Understanding these relationships should have a positive carry-over effect in their personal lives. As children develop a sense of responsibility and commitment toward other living creatures, they will apply these attitudes to the human condition.

Classroom Activities

Classroom activities with live organisms can center on behavioral, morphological, ecological, or other emphasis categories, and involve the practice of basic science skills such as measuring, observing, and collecting and analyzing data.

Suggested general classroom activities include:
• Learn to care for life by providing suitable habitats, food, water, and other needs
• Observe differences between plants and animals
• Study life cycles
• Observe growth, development, and reproduction
• Observe food chains and webs
• Study behavioral responses to normal ranges of environmental factors
• Observe differences among individuals of the same species
• Measure growth responses to environmental factors
• Study population growth
• Culture algae, protozoa, and invertebrates
• Clone plants (duckweed, African violet, etc.)
These investigations are suitable for open-ended assignments and science fair projects.

Observational studies also can be done with living organisms. Many animal behaviors are easy to identify and record. Behaviors also vary between individual of the same species and individuals of different species, providing grounds for comparison and contrast. Observations can be made in the following kinds of animal behaviors:

Table 1
Concepts used in *Classroom Creature Culture*

Activities Using Living Organisms:
  A Big Lesson in A Small Pond, Cryptozoa, Duckweed, Mealworms in the Classroom, Reining Monarchs, Planaria, Seed Plants

Adaptations:
  Green Anole (American Chameleon), Praying Mantises, Walking Sticks

Animals in the Classroom (Proper Care and Treatment):
  NSTA Position Statement

Collecting:
  Collecting and Observing Algae, Collecting and Observing Protozoa, Daphnia, Newts

Culturing Techniques:
  Culturing Protozoa, Growing Algae in the Classroom, Recycling Plastic, Reining Monarchs, The Establishment of a Life Science Culture Center, Seed Plants

Habitats:
  A Big Lesson in A Small Pond, A Freshwater Aquarium, A Saltwater Aquarium, Cryptozoa, Daddy Longlegs, Anacharis, Newts, Snakes As Pets, Terrestrial Isopods

Life Cycles:
  Daddy Longlegs, Mealworms in The Classroom, Newts, Reigning Monarchs, Walking Sticks

Natural History:
  Daddy Longlegs, Newts, Reining Monarchs, Tiger Salamanders

Morphology and Structure:
  Daddy Longlegs, Daphnia, Duckweed, Newts

Observing:
  Daphnia, Mealworms in the Classroom, Daddy Longlegs, Newts

- Locomotion in animals
- Nesting habits and care of young
- Courtship and mating
- Schooling behavior in fish
- Trail marking and food preferences in ants
- Territoriality and communication behavior in crickets
- Maze learning in small mammals (rewards only)
- Construction of observation chambers (beehives, ant colonies, spider webs, wasp nests)

Organism Care

Before any living organism-plant or animal can be kept in captivity, certain precautions *must* be taken.
- Do not try to keep too many different species in the classroom at the same time. Too many organisms may exhaust your resources and time.
- Always maintain animals in clean quarters.
- Provide adequate habitats for animals or containers for plants before bringing them into the classroom.
- Keep only organisms that can tolerate normal classroom conditions and require little effort to maintain.
- Provide materials necessary for the maintenance of organisms in a convenient location.
- If a wild animal is brought into the classroom, keep it for a short time and release it where it was found.
- Do not collect wild birds or mammals. These animals can carry human diseases. They also can suffer the greatest harm in captivity.
- For classroom observations of animals, always prefer invertebrates. They are easier to maintain.
- Do not take an organism from its natural habitat if it is an endangered species.
- Do not keep organisms that are dangerous to handle.
- Never force students to handle anything that they find objectionable or fearsome. Direct willing students to observe without touching first, then to touch and later hold. Animal and student need time to become accustomed to each other.
- Remember, teacher behavior will become the standard to imitate. The difference between experimentation with and cruelty to animals is a fine one. Students below the college level have neither the background experience nor the judgment to undertake complicated experiments with animals. *At all times, children should be taught to treat animals kindly.* The use of living animals should be limited to observation of normal living patterns and should never take an organism beyond its range of tolerance to environmental conditions. Experimental procedures with animals that involve nutrient deficient diets, discomfort, pain, or death should never be attempted. For more detailed humane animal treatment guidelines, see page 11 of this publication.

An excellent source of information about endangered species and some teaching aids on animals is the *Endangered Species Handbook*, published by the Animal Welfare Institute. For a compilation of teaching strategies integrating wildlife-related studies with various subjects in the curriculum, contact a representative of your state's wildlife management agency about Project WILD. Also ask about staff development workshops in your local school system. Table 2 contains a partial list of agencies that provide teachers with free or low-cost educational materials on plants, animals, or environmental concerns.

This introduction was written especially for this new edition by Carolyn H. Hampton.

Table 2
Sources for Teaching about Plants, Animals, and the Environment*

American Forest Institute
Education Division
1619 Massachusetts Avenue
Washington, DC 20036

Animal Welfare Institute
P.O. Box 3650
Washington, DC 20007

Forest Service, USDA
P.O. Box 2895
Washington, DC 20013

Center for Action on Endangered Species
National Audubon Society
175 West Main Street
Ayer, MA 01432

Center for Environmental Education
1925 K St., N.W., Suite 206
Washington, DC 20006

Florida Conservation Foundation, Inc.
Environmental Information Center
935 Orange Avenue
Winter Park, FL 32789

National Audubon Society
1130 Fifth Avenue
New York, NY 10028

Geographic School Bulletins
School Service Division
National Geographic Society
P.O. Box 2895
Washington, DC 20013

U. S. Dept. of the Interior
Information Office
Interior Building
Washington, DC 20242

National Wildlife Federation
1412 16th Street, N.W.
Washington, DC 20036

Project WILD
Salina Star Route
Boulder, CO 80302

Woodsy Owl
Forest Service, USDA
P.O. Box 2417
Washington, DC 20013

St. Regis Paper Company
150 East 42nd Street
New York, NY 10017

Walt Disney Educational
 Media
500 S. Buena Vista Street
Burbank, CA 91521

U. S. Dept. of Agriculture
Office of Information
Washington, DC 20250

* Write for lists of available educational materials on school letterhead.

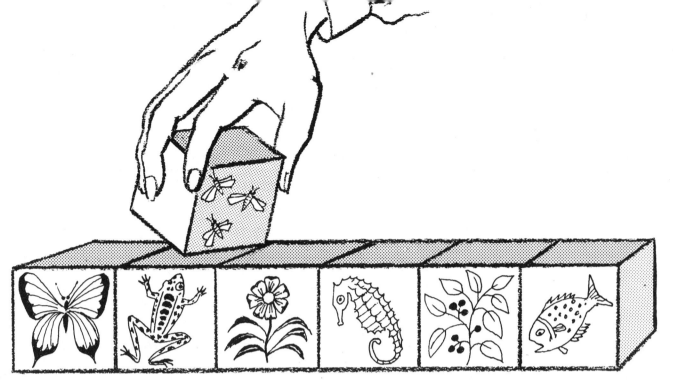

MARY VILLAREJO

The Establishment of a Life Science Culture Center

NATIONAL SCIENCE FOUNDATION (NSF)-supported elementary science curriculum programs nationwide have significantly increased the use of living organisms in the classroom. A need arose to aid schools in procuring living organisms for reasonable expenditures, and to increase the effort in preservice training of elementary school teachers in the culturing and maintenance of living organisms. The Department of Science Education at East Carolina University has provided for both needs by setting up a life science culture center.

In 1972, at the close of a summer institute to train elementary teachers to implement the SCIS program, there were more living organisms than on the day the large shipment arrived. A vacated physics research laboratory was transformed into a life science "home" for cultures. The room was not particularly suited for the use to which it was applied, but there were shelves, counters, some cabinets with doors, one wall of windows facing west, and good fluorescent lighting. No renovations were made. A three-tiered plant mobile and a mobile shelf unit were moved into the end of the room near the windows. All of the extra aquaria and terraria in the department were brought to the center, repaired, and made water tight. Some were modified by replacing the broken glass with screen wire. These made excellent cages for crickets and chameleons.

Hard work and creative scrounging produced a culture center at no cost.

A large collection of wide-mouth gallon jars were to serve as homes for mealworms, algae, daphnia, et al. Baby food jars were rounded up also for the life science culture center. The cost was zero—plus some hard work in scrounging or repairing.

The authors had been teaching a course for preservice elementary

Reprinted from *Science and Children*, Carol D. and Carolyn H. Hampton, April 1978, pp. 7-11.

education majors entitled "Investigations in Biological Science." This course consisted of several units of study that would be particularly helpful to elementary science teachers. It was modified immediately to include a unit on the culturing and maintenance of living organisms for the elementary classroom.

Students took on the tasks of rearing specific organisms, researching the literature for techniques, trying out techniques, and reporting results. Some techniques were abandoned when results were not satisfactory. If two different techniques were equally successful, the simpler one was retained. In time, we had compiled a file index which listed culturing techniques selected by the following criteria: (1) simplicity, (2) ease of carrying out, (3) low expense, and (4) availability of materials.

Since the establishment of the center, students enrolled in the investigations course have continued to study culturing techniques. (Later students in the Federal assistance programs with course experience were given assignments in the culture center.) We make cultures available free of charge to classrooms of nearby schools. To obtain free cultures a teacher has only to: (1) notify the center one week in advance, (2) bring baby food jars to replace those in which they receive their cultures, and (3) have the cultures picked up before 4:30 p.m. on the designated date. A checklist of available organisms provides a convenient order form.

A booklet, *Living Organisms for the Elementary Classroom,* was compiled for a reference manual that the students could keep. The State Department of Public Instruction has made this booklet available free of charge to all elementary schools in the state.

In the summer of 1975, NSF funded a three-week program to bring teams of three persons from various school systems in North Carolina to the East Carolina University campus to learn culturing techniques and to develop plans for setting up culture centers in their respective school systems. The typical group consisted of a school supervisor, a high school biology teacher, and an elementary school teacher. The supervisor represented school administrative support; the

high school teacher with life science expertise would involve high school students in maintaining a center; and the elementary school teacher knowledgeable about child development would act as liaison between the team and other elementary teachers in the school system. At the end of the program, each team was prepared to go back to its respective school and set up culture centers modeled after ours. (In one case a mobile trailer was used as there were no vacant classrooms.) In addition, they could conduct staff development workshops to prepare others to implement the program (in our case, SCIS).*

When news about the center spread, we began to receive calls from inservice teachers, inquiring about the possibility of a course in culturing. Consequently, we have developed a unit on the maintenance of living organisms for incorporation in a graduate level science education course for practicing elementary teachers.

One of the most rewarding outcomes of the Life Science Culture

*Science programs besides the Science Curriculum Improvement Study (SCIS) which utilize living organisms are the Ginn Science Program, Focus on Science Series, Elementary Science Study, and The Elementary School Sciences Program.

Center has been the interrelationships and feedback mechanisms that have evolved from the interaction between the center, public school classrooms, the university's program of science education for preservice and inservice teachers, and the State Department of Public Instruction.

Design Your Own Center

For individual schools or school systems which would like to attempt the maintenance and operation of their own life science culture centers we have outlined a basic plan. First, keep in mind the achievement of the following objectives:

1. To keep commonly used organisms readily available in large numbers and in healthy conditions. Many of the headaches in working with living organisms in the classroom stem from shipping, i.e., delayed or lost shipments accounting for long waits with organisms living under less than optimum conditions. Once a population has been procured and is thriving, further shipping costs are saved.

2. To take advantage of rapid population growth of certain organisms to supply large numbers of classrooms. Many organisms can be

FIGURE 1: MOBILE UNITS FOR ORGANISMS

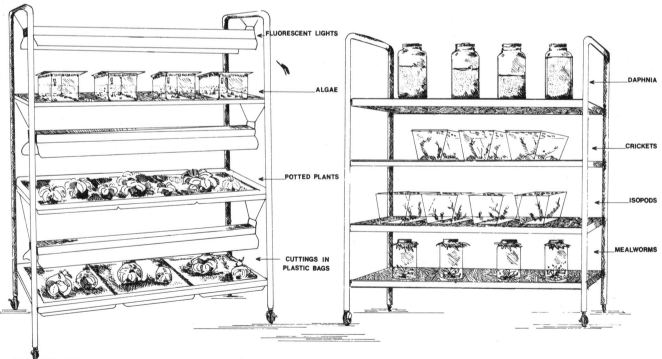

FLUORESCENT LIGHTS
ALGAE
POTTED PLANTS
CUTTINGS IN PLASTIC BAGS
DAPHNIA
CRICKETS
ISOPODS
MEALWORMS

ARTWORK BY EAST CAROLINA UNIVERSITY

grown with simple and inexpensive procedures. Beyond the initial cost of establishing the population, there remains virtually little or no expense in maintaining the organisms for the remainder of the school year. Examples are mealworms, isopods, and crickets.

3. **To maintain larger expensive organisms that do not reproduce well in captivity, e.g., frogs, chameleons, for use in classrooms on a loan basis.** Providing some of the larger organisms to each classroom for certain activities can be expensive. By careful scheduling, organisms can be made available to blocks of classes for a semester, a quarter, or even shorter periods of time. Healthy organisms can be scheduled over and over again.

Helping in the operation of a school-maintained culture center can provide the following opportunities for students:

1. Participation in individualized or small group learning experiences,
2. Development of library research and index skills,
3. Assumption of responsibility for the care and maintenance of specific organisms, and
4. Inculcation of many fundamental principles of life science and ecology.

Living Culture Sources

Initial cultures of living organisms may be obtained from commercial supply houses. Organisms requiring a minimum of time, space, and expense are listed in the box. Many organisms may be obtained free from non-commercial sources. (See Table I.) Enlist students in the collection of organisms for the center whenever possible.

Since all organisms living in a very limited space use up their food supply, overpopulate, and accumulate toxic wastes or offensive products fairly rapidly, all cultures must be routinely fed, cleaned, and transferred. Students can be trained to carry out routine maintenance procedures.

Equipment and Maintenance Requirements

Water

Always use unchlorinated water. Spring water and rainwater are the best sources. In many cases, tap water can be used if allowed to "stand" uncovered for three days to allow the gaseous chlorine to

---

Easy Care Cultures

Invertebrates: hydra, isopods, pond snails, land snails, crickets, mealworms, brine shrimp, fruit flies, daphnia, wingless pea aphids.

Vertebrates: guppies.

Algae: chlamydomonas, euglena.

Aquatic Plants: elodea (Anacharis), eelgrass, duckweed.

Potted Plants: geranium, coleus.

TABLE 1. Non-Commercial Sources and Containers for Organisms

Organisms	Non-Commercial Sources	Culture Containers
POND SNAILS	Fresh water ponds, creeks	Aquaria, large battery jars, gallon glass jars
LAND SNAILS	Mature hardwood forests: on rocks, fallen logs, damp foliage	Terraria, large battery jars
ISOPODS AND CRICKETS	Under rocks, bricks, and boards that have lain on ground for some time; between grass and base of brick buildings	Glass or plastic terraria, plastic sweater boxes (Provide vents in cover.)
MEALWORM BEETLES	Corn cribs, around granaries	Gallon glass jars with cheesecloth covers
FRUIT FLIES	Trap with bananas or apple slices. (Place fruit in a jar with a funnel for a top.)	Tall baby food jars, plastic vials (Punch holes in jar lids, cover with masking tape and then prick holes in tape with a pin.)
WINGLESS PEA APHIDS*	Search on garden vegetables, e.g., English peas.	On pea plants potted in plastic pots, milk cartons (Keep aphids in a large terrarium so they cannot wander to other plants in the school.)
GUPPIES	Obtain free from persons who raise guppies as a hobby (usually glad to reduce population when they clean tanks.)	Aquaria, large battery jars
CHAMELEONS*	Dense foliage along river banks or railroad tracks (Catch with net or large tea strainer.)	Prepare cage from broken aquaria. (Broken glass can be replaced by taping cloth screening along sides.) See Figure 3.
FROGS*	Along edges of ponds, ditches, creeks (Catch with large scoop net.)	Large plastic ice chest (Set near a sink so a constant water supply can be provided.) See Figure 2.
CHLAMYDOMONAS AND EUGLENA	Freshwater ponds	Gallon glass jars, aquaria, battery jars
ELODEA (ANACHARIS)*	Ponds, creeks; usually along edge or in shallows	Aquaria, large battery jars
EELGRASS*	Wading zone of brackish water	Aquaria, large battery jars
DUCKWEED	Edge of ponds or fresh water swamps	Aquaria, large battery jars
COLEUS AND GERANIUM	Persons who raise them (Start by rooting cuttings in 1 part sand, 1 part vermiculite, in plastic bags; keep moist.)	Clay pots, milk cartons, tin cans

*These species are difficult to obtain from their natural habitats. Unless you have a convenient source, it is better to buy them commercially. Try a local aquarium or pet shop.

escape before being used. A suggested way of maintaining a ready source of "aged" tap water is to keep five 5-gallon buckets of aged water available. When a bucket is emptied, refill it with fresh tap water; label the date with masking tape; let it stand uncovered for three days; then cover it to prevent further evaporation and contamination. Always use the water marked with the oldest date.

Temperature

The temperature of the culture center should be kept between 70° and 75 °F (21-25 °C). Since guppies reproduce best at temperatures between 75° and 85 °F, aquarium heaters can be used to warm the guppy containers.

Avoid too much heat. High temperatures lower the oxygen concentration in water and increase the metabolic activity of organisms and thus increase their rate of oxygen consumption.

Avoid rapid changes in temperature. When fish arrive, place them (with water they had come in) in a small jar or plastic bag in the tank about 30 minutes allowing the water temperature in the jar or bag to become the same as that in the aquarium. Then, carefully tilt the jar or bag to allow the fish to swim into the tank.

Glassware

All glassware must be thoroughly cleaned and rinsed. Gallon glass or plastic jars may be obtained from cafeterias and restaurants. Students can usually supply baby food jars from home.

Lids

Unless culture methods specify otherwise, maintain a loose cover over all cultures to prevent contamination with dust and unwanted microorganisms. Every aquarium should be equipped with a loose glass cover. Glass covers can be custom-made, but be sure the edges are sanded or polished to prevent cuts.

Physical Space and Lighting

Maintain cultures in an area where each receives proper lighting. Plants should be placed under fluorescent lights or near a window

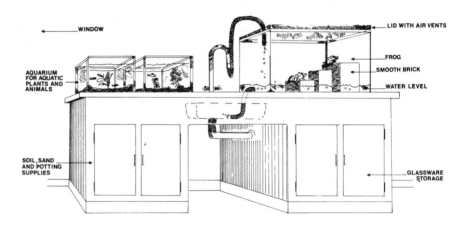

FIGURE 2: ARRANGEMENT OF AQUARIUM AND FROG TANK

exposed to full sunlight at least part of the day. Small animals can be placed in minimal light. Keep cultures stationary, well labeled, and in full view. (See Figures 1 and 2.)

Provide storage space for subculturing equipment, cleaning, and food supplies. Labeled shoe boxes or plastic sweater boxes, one or more for each kind of organism work well. (See Figure 3.) Also, you will need a work area for new cultures. Perhaps the maintenance department or the industrial arts department of your school system can build a simple table or window shelf.

Supplies

If a culture center is to function smoothly, a well-balanced stock of supplies should be maintained. Food for organisms, media materials, containers for housing organisms, and maintenance supplies should be readily available at all times. (See Figure 3.) The most basic and necessary equipment for maintaining the organisms listed in this article includes:

General Supplies: containers, lids, aquarium sand, potting soil, noniodized salt (brine shrimp cultures), and aged tap water.

Maintenance Supplies: basters (transfer small organisms), buckets, coat hangers (dip net handles), dip nets, old stockings and nylon slips (nets and straining), paper towels, sponges, old rags, sprinkler bottles,

toolbox (hammer, nails, pliers), and trowels (potting plants).

Food for Organisms: apples (crickets), drosophila medium, fish food, meal—bran flakes and oatmeal—(mealworms), potatoes (isopods), and rye grass (crickets).

Culturing Methods

Space does not permit the authors to give culturing techniques for each organism. However, there are a large number of references listed at the end of this article which have been selected for their clarity and ease of reading. For a novice, and particularly for an elementary school program, one of the best sources is *A Sourcebook for the Biological Sciences* by Morholt, Brandwein, and Joseph.

References

1. Behringer, Marjorie P. *Techniques and Materials in Biology.* McGraw-Hill, Inc., New York City. 1973.
2. Flagg, Raymond. *Carolina Drosophila Manual.* Carolina Biological Supply Company, Burlington, North Carolina. 1973.
3. Hooft, Jan, and Robert F. Bayly. *Care of Living Plants.* Carolina Biological Supply Company, Burlington, North Carolina. 1975.
4. James, Daniel E. *Carolina Marine Aquaria.* Carolina Biological Supply Company, Burlington, North Carolina. 1973.
5. James, Daniel E. *Culturing Algae.* Carolina Biological Supply Company, Burlington, North Carolina. 1974.

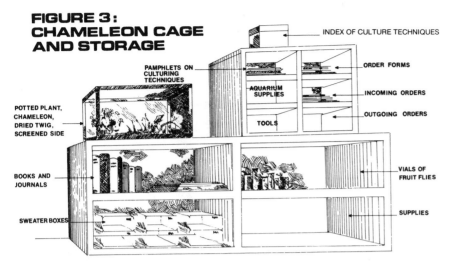

6. James, Daniel E. *Fungi Exercises. Culture and Techniques for the Introductory Botany Course.* Carolina Biological Supply Company, Burlington, North Carolina. 1974.

7. Miller, David F., and Glenn W. Blaydes. *Methods and Materials for Teaching the Biological Sciences.* McGraw-Hill, Inc., New York City. Second Edition. 1962.

8. Morholt, Evelyn, et al. *A Source-book for the Biological Sciences,* Second Edition. Harcourt, Brace and World, New York City. 1966.

9. Needham, James G. *Culture Methods for Invertebrate Animals.* Dover Publications, New York City. 1937.

10. Orlans, F. Barbara. *Animal Care From Protozoa to Small Mammals.* Addison-Wesley Publishing Co., Reading, Massachusetts. 1977.

11. Pennak, Robert W. *Fresh-Water Invertebrates of the United States.* Ronald Press, New York City. 1953.

12. Perkins, Kenneth W., and Richard L. Franks. *Amphibian Culture.* Carolina Biological Supply Company, Burlington, North Carolina. 1976.

13. Turtox Service Leaflets. *Raising Live Insects in the School Lab* (#7), *Flowering Plants in the Lab* (#8), *The School Terrarium* (#9), *The Culture and Lab Use of Planaria* (#10), *Getting to Know the Brine Shrimp* (#11), *How to Feed Animals in Aquaria and Terraria* (#12), *Aquatic Insects in the Lab* (#14), *How to Culture Drosophila and Demonstrate the Laws of Heredity* (#15), *Presenting: The Vinegar Eel as a Study Specimen* (#16), *Caring for Rats, Hamsters, Guinea-Pigs, and Gerbils* (#17), *Establishing and Maintaining Marine Aquaria* (#20). Turtox/Cambosco, Chicago, Illinois. 1974.

14. Whitten, Richard H. *Care of Living Invertebrates.* Carolina Biological Supply Company, Burlington, North Carolina. 1976.

NSTA Position Statement

Guidelines for Responsible Use of Animals in the Classroom

These guidelines are recommended by the National Science Teachers Association for use by science educators and students. They apply, in particular, to the use of nonhuman animals in instructional activities planned or supervised by teachers who teach science at the precollege level.

Observation and experimentation with living organisms give students unique perspectives of life processes that are not provided by other modes of instruction. Studying animals in the classroom enables students to develop skills of observation and comparison, a sense of stewardship, and an appreciation for the unity, interrelationships, and complexity of life. This study, however, requires appropriate, humane care of the organism. Teachers are expected to be knowledgeable about the proper care of organisms under study and the safety of their students.

- Acquisition and care of animals must be appropriate to the species.
- Student classwork and science projects involving animals must be under the supervision of a science teacher or other trained professional.
- Teachers sponsoring or supervising the use of animals in instructional activities—including acquisition, care, and disposition—will adhere to local, state, and national laws, policies, and regulations regarding the organisms.
- Teachers must instruct students on safety precautions for handling live animals or animal specimens.
- Plans for the future care or disposition of animals at the conclusion of the study must be developed and implemented.
- Laboratory and dissection activities must be conducted with consideration and appreciation for the organism.
- Laboratory and dissection activities must be conducted in a clean and organized work space with care and laboratory precision.
- Laboratory and dissection activities must be based on carefully planned objectives.
- Laboratory and dissection objectives must be appropriate to the maturity level of the student.
- Student views or beliefs sensitive to dissection must be considered; the teacher will respond appropriately.

—Adopted by the NSTA Board of Directors in July, 1991

Reprinted from *Science and Children*, Carol D. and
Carolyn H. Hampton, September 1979, pp. 50-52.

Collecting and

If microscopes are available, the study of algae will bring enjoyment and learning to students and satisfaction to teachers. Seeing the intricate beauty and the natural appearance of plants that are relatively foreign to students' experiences can be a revelation to middle and junior high school pupils.

Freshwater algae are widely distributed and easily collected. Algae can be found growing in moist places—every type of damp and aquatic habitat—in roadside ditches, bird baths and animal watering troughs; on tree bark, damp soil and rocks, and the backs of turtles; and on ponds as floating mats and tufts, or attached to the substratum of ponds, lakes, and flowing streams.

Many species of algae occur throughout North America because vegetative forms are transported from one body of water to another on the feathers and legs of waterfowl, and spores are borne by the wind. Algae's occurrence depends on a combination of environmental factors such as light penetration and intensity, and the temperature, chemical composition and pH of the water, rather than geographic locations.

Collection

You will need the following equipment:

6 to 8 glass collecting jars (baby food
jars with lids)
6 plastic bags
large plastic bucket
shallow plastic dishpan (white)
fine-mesh dip net or tea strainer
long pipette or baster
plankton net
trowel or spatula
pocket knife
hand magnifier
clipboard and paper
labels or masking tape

The best times of the year for collecting are spring and fall. Populations of algae decrease during the summer and winter due to temperature extremes. During spring or fall, classroom temperatures are more easily adjusted for keeping samples several days or for trying to culture algae. High temperatures are unfavorable for most algae.

The best times to collect algae are the spring and fall, when temperatures are mild.

A plastic tote tray with a handle, usually available in variety stores, makes a handy field tray for collecting. Stock the tray with baby food jars and lids, plastic bags, a pocket knife, a long pipette, a fine mesh-dip net, hand lens, labels, and a pencil.

Put each sample in a separate container and label with code numbers. Make notes on the clipboard about location and growth. This information will be helpful in the classroom when you identify the algae.

Don't fill the jars more than a quarter full with algae. Add water from the habitat to fill the containers half full. Leave plenty of room for air in the containers while transporting the collection back to the classroom. You can bring back additional water from the habitat in a large bucket. As soon as possible uncap the lids to prevent decay and deterioration.

Terrestrial Algae. Use a trowel or spatula to scoop up soil with green, or purplish to black colored patches. Put the material in a plastic bag and label it. Algae growing on damp rocks, bricks, or clay flower pots may be scraped with a pocket knife into a small jar. Collect algae growing on the bark of trees by picking off several pieces of bark.

Aquatic Algae. With a knife blade you can scrape or pry loose encrusting filamentous algae, catching the samples in a fine-mesh dip net as they fall. Algae that form a scum, gelatinous mass, or filamentous mats that float on surface water may be skimmed from the surface

Figure 1. Homemade Plankton Net

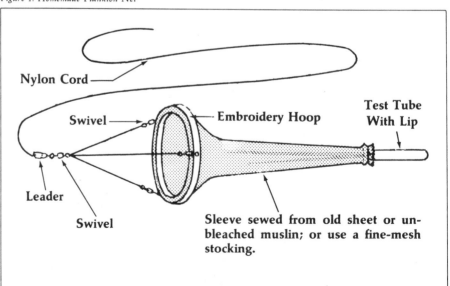

Nylon Cord

Swivel — Embroidery Hoop

Test Tube With Lip

Leader

Swivel

Sleeve sewed from old sheet or unbleached muslin; or use a fine-mesh stocking.

Observing Algae

by a fine-mesh net or kitchen strainer.

Planktonic—free-floating—forms may turn up in mixed collections, but are best concentrated for classroom study by using a plankton net. (See Figure 1.)

The cord of the plankton net (which should be 2 to 3 meters in length) may be attached to a broom handle. A student can walk along the bank of the pond or lake, trailing the plankton net at or just below the water's surface. A second method uses a long cord. One student standing on the bank holds the end of the cord while a second student walks with the net, around the edge of the pond. Be careful not to tangle the cord in the vegetation. Upon a signal the second student throws the net into the water. The first student reels in the cord rapidly. Your collecting technique should keep the net slightly under the surface of the water. If the cord is pulled in too fast the net will skip over the surface of the water; if pulled in too slowly, the net will sink and drag in the bottom mud.

Without a net, concentrated collections of desmids, diatoms, and planktonic forms may be collected by squeezing algae mats and aquatic seed plants. Hold handfuls of material over a wide-mouth container. Squeeze until nothing more drips into the container. "Squeezings" of sphagnum moss, bladderworts, and water milfoil usually contain desmids and diatoms.

Forms that grow attached to plants may be collected by gathering rooted aquatic plants. Children may wade into the water to gather these or they may drag them to shore using a plant grapple. (See Figure 2.)

Throw out the plant grapple into the zone of submerged vegetation. Pull the grapple toward the shore. As soon as the plants are hauled in, put them in a plastic dishpan half filled with clear water from the habitat. Individual stems and plant parts may be "stripped" between the thumb and forefinger over collecting containers.

Algae forms that live on the bottom mud can be collected by a long pipette or a commercial poultry baster. The pipette may be constructed by attaching a rubber bulb to a 30 to 35 cm length of glass tubing. Squeeze the rubber bulb, then move the open end of the glass tube slowly over the bottom surface while gradually relaxing pressure on the bulb. When the bulb is relaxed, squeeze the tube's contents into a collecting jar.

Algae that form mats and gelatinous scums on submerged rocks, stream banks, or on the walls of dams, under stones in flowing water, and sides of waterfalls can be scraped free with a knife blade and caught in collecting jars.

Observation

You will need the following equipment:

microscopes
microscope slides
deepwell slides
flat toothpicks
glass coverslips
droppers
lens paper
cotton fibers
petroleum jelly
paper towels

After returning to the classroom, pour collections containing organisms into wide shallow plastic or glass dishes. You can keep filamentous forms in large glass jars in a cool place with reduced illumination. A window facing north affords the best light conditions. Observe the planktonic forms, "squeezings," and "strippings" as soon as possible as they deteriorate rapidly. Terrestrial forms can usually be kept in a container lined with wet paper towels. Flood dried scum with water for a few hours before pipetting drops for slide preparation. You can keep many filamentous forms in a refrigerator for several days.

Prepare wet mounts of water containing algae or put a few strands of filamentous algae on a slide with a drop of water. Hold the coverslip at a 45° angle with the slide. Let it fall gently onto the drop. Put the slide under the microscope, observing first under low power. Organisms that remain stationary may be observed under high power. If there are motile forms or zoospores, prepare

Figure 2. Plant Grapple

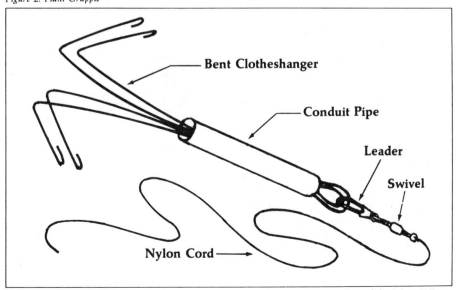

Artwork by Mary Villarejo

Algae Groups

Characteristics	Blue-Green Algae	Green Algae	Euglenoids	Dinoflagellates	Diatoms
Color	Blue-green	Grass-green	Green; some colorless	Yellowish to reddish brown; some colorless	Yellow-green to golden brown
Pigment Location	Throughout the cell	In plastids	In plastids	In plastids	In plastids
Cell Wall	Cannot distinguish from slimy coating	Semi-rigid, smooth, or with spines	Absent; changes in body shape called euglenoid movement	Resembles plates of armor	Very rigid. Made of silica—two halves fit together like a pill box, finely sculptured
Nucleus	Absent	Present	Present	Present	Present
Flagellum	Absent	Absent	Present	Present; one trails behind, other whips in a groove around cell	Absent
Eyespot	Absent	Absent in most; present in some single-celled or colonial forms	Present	Absent	Absent
Form	Occur as single cells, filaments, or colonies surrounded by slimy sheaths	Occur as single-cells, filaments, flattened, or spherical colonies; most lack slimy sheaths	Single-cells	Single-cells	Single-cells or ribbon-shaped colonies

another slide. Put a few cotton fibers on the slide. Organisms will be trapped in the small spaces enclosed by the fibers.

You can best study colonial and motile forms by the hanging drop method. First prepare slide wells. Apply clear fingernail polish to a clean glass slide forming a ring that will fit underneath a coverslip. Allow the ring to dry before applying the polish again. Continue to build up successive layers until a small well about one millimeter in depth is formed. Prepare several well slides and let them dry thoroughly.

To make a hanging drop, carefully apply a film of petroleum jelly around the top rim of the fingernail polish well with a flat toothpick. Don't get petroleum jelly inside the well. Put a drop of water containing motile forms in the center of a clear glass coverslip. Invert the slide, centering the well over the drop of water on the coverslip. Press gently to seal the well rim to the cover-

slip. Turn the slide over with the coverslip up to put under the microscope. Don't run the microscope objective through the coverslip. The preparation should last for several hours without drying out.

Put a container with "squeezings" under a light source. Place a black sheet of construction paper over one half of the container. Be sure that the light source does not heat the water. With a dropper, pipette water from the light side and the dark side of the container. Prepare slides of the water from both regions. Which organisms are positive phototaxic—move toward the light—or negative phototaxic—move away from the light? In checking other samples that have been left undisturbed for a while be sure to take water from near the surface and at the bottom of the containers. Some forms are free-floating while others reside on the bottom substrate.

An excellent activity for studying dif-

ferent pond habitats and descriptions for constructing homemade collecting devices are found in "Habitats of the Pond," *OBIS, Trial Edition I.* A handy and inexpensive reference for identifying algae and other pond organisms is the Golden Nature Guide, *Pond Life.*

References
Smith, Gilbert M. *The Freshwater Algae of the United States.* Revised Edition. McGraw-Hill Book Company, New York City. 1950.

Identification Aids
Hausman, Leon A. *Beginner's Guide to Freshwater Life.* G. P. Putnam's Sons, New York City. 1950.
Klots, Elsie B. *The New Field Book of Freshwater Life.* G. P. Putnam's Sons, New York City. 1966.
Reid, George K. *Pond Life.* Golden Press, New York City. 1967.

Growing Algae in the Classroom

Teachers can use freshwater algae in middle grade and junior high school classroom experiments to demonstrate basic biological concepts including requirements of living organisms. Teachers may not use living algae cultures because the cultures can be expensive. They may think culturing is impractical. While some algae are impractical for classroom use, many forms can be maintained with few technical requirements. In fact, most algal forms useful in classrooms can be cultured using short-term methods described here.

You do not need pure culture techniques for keeping algae for class use. If you monitor temperature, light, pH, water quality, and have a satisfactory culture medium, algae are easy to grow in your classroom.

Culturing algae will make students aware of the basic requirements of living organisms.

Temperature. Many species of algae will keep at room temperature. The optimal range of temperatures for classroom culturing is between 10 °C and 21 °C. Avoid temperatures above 27 °C because higher temperatures are more damaging than lower ones.

Light. Keep cultures in a well-lit room near a window with a northern exposure. *Never* put algae in direct sunlight because the water may heat beyond the

Reprinted from *Science and Children*, Carol D. and Carolyn H. Hampton, February 1980, pp. 40-41.

plant's tolerance. An ideal set-up is to install a 45-watt fluorescent tube 30 to 45 cm above the storage shelf. Between 200 to 400 footcandles of light will result, adequate light for many species. You can put an inexpensive timer between the wall socket and the light plug and set it to furnish 16 hours of illumination and 8 hours of darkness.

pH. Keep the hydrogen ion in concentration at neutral to slightly alkaline (pH of 7 to 7.5). Check your water source with pH test paper or have the water tested by a high school chemistry teacher who has access to a pH meter to make sure the pH is correct.

Water. The quality of water is of critical importance. Use filtered pond water, spring water, rainwater, or in some cases, distilled water. *Never* use chlorinated tap water. Heat all water for making solutions to 73 °C for at least 20 minutes and let it cool to room temperature before inoculating the water with organisms. Refer to specific formulas for media to see whether or not water should be made aseptic before or after mixing with other materials. You can make aseptic water by heating water in a covered container such as a stainless steel stockpot. Store the water in its covered pot and keep it in a place where it will not be disturbed.

Mark the water level on the side of the culture vessel and compensate for evaporation by adding distilled water, aseptic spring, or rainwater. A glass poultry baster makes an excellent device for adding aseptic water to culture vessels. Keep the baster clean by storing it in a graduated cylinder with a beaker inverted over the bulb. Keep culture vessels covered with loose fitting lids or plugs to allow gas exchange and to keep out unwanted microorganisms and other forms of algae.

Glassware, Workspace, and Storage Preparation. Carefully wash all glassware used in culturing with warm water and detergent. Rinse twice in warm tap water. Residual detergent on glassware may kill algae so be sure the glassware is carefully rinsed. Store glassware upside down in a clean area. Sterilize pipettes or droppers used in transferring algae by boiling them in water for 10 minutes. Store the sterile pipettes or droppers, tips down, in a wide-mouth canning jar with a tightened lid. *Do not* sterilize rubber bulbs. You can keep these in a box until time for making transfers.

Scrub working and storage surface with a two percent lysol solution made of 2 mL of Lysol® and 98 mL of water to help prevent contamination of cultures with unwanted algae and other microorganisms.

Culture Media

The following media have proven successful with culturing certain species of algae under average classroom conditions.

Fertilizer Medium. Mix together 1 g of 4-10-4 or 5-10-5 (nitrogen, phosphorous, potash) fertilizer and 1 L of pond, spring, or rainwater. Heat the mixture to 30 °C for 20 minutes. Filter the mixture while still hot, before cooling it and inoculating the water with algae. Use for *Eudorina, Pandorina, Volvox,* and other volvocales.

Pringheim's Soil-Water Medium. This medium represents a miniature pond. The soil provides organic matter, growth factors, and trace elements. We have found that gas collecting bottles fitted with cotton plugs or heat resistant plastic foam plugs make satisfactory containers. Put a pinch of calcium carbonate, $CaCO_3$, in the bottom of the bottle. Layer over this 1.5 cm of rich loamy garden soil. Soil you have recycled from potted plants or greenhouses works well in this method. Fill the bottles to the shoulder with rain, spring, or distilled water. Plug with cotton or plastic foam stoppers. Put the bottles in a water bath and steam them at 70 °C for 15 minutes on two consecutive days to kill both vegetative and spore forms of algae and fungi that are in the soil. Let the bottles cool before inoculating. Use for Blue-green Algae: *Oscillatoria*, *Rivularia*; Green Algae: *Chalmydomonas*, *Closterium*; other Desmids, *Eudorina*, *Pandorina*, *Volvox*, *Cladophora*, *Oedogonium*, *Spirogyra* (omit $CaCO_3$) and *Zygnema*.

Natural Water Medium. Fill a clean 3.8 L jar with water from the source where algae is found growing. Do not crowd the filamentous forms. Set the container in a north window. Use for *Rivularia*, *Chlorella*, and others for a few days.

Natural Pond Medium with Mud/Sand Substrate. This method also represents a miniature pond and is useful for some of the larger branching filamentous algae, particularly the stoneworts. Use a wide-mouth 3.8 L jar or a battery jar. Layer 1.5 cm of pond mud and/or sand. Cover with 1.5 cm layer of prewashed aquarium gravel. Fill the container with filtered pond water without disturbing the bottom layers. Allow sediment to settle before planting the filaments in the substrate, using gravel to hold them in the bottom. Use for Stoneworts: *Chara* and *Nitella*.

Aquarium Method. Keep a classroom aquarium or battery jar stocked with a few rooted plants and two or three small fish. You can introduce filamentous algae for short intervals of time. Clean away unwanted algae by hand. Use for Blue-green Algae: *Nostoc*, *Oscillatoria*, *Rivularia*; Green Algae: *Hydrodictyon*, *Scendesmus*, Desmids, *Cladophora*, and diatoms. *Spirogyra* and *Zygnema* can also be used although neither form may live a long time.

Wood Ash Medium. Obtain hardwood ashes from a fireplace, for instance, and sift them. Layer the sifted ashes about 3 to 5 mm deep in the bottom half of a petri dish. Moisten the ashes with distilled or aseptic spring water, or rainwater. Inoculate the ashes with Blue-green Algae, cover with a lid, and put in the light. Use for Blue-green Algae: *Anabaena* and *Gleocapsa*.

Wood Bark Method. Keep pieces of bark on which algae is found growing in a chamber such as a petri dish. Keep the bark moist with aseptic spring or rain-water. Use for *Pleurococcus* (formerly *Protococcus*).

Transferring Cultures

Moving algae from one container to another is called "transferring." For classroom short-term cultures, algae should be transferred every two to four weeks, depending on growth rate of individual species. Take care in transferring procedures to avoid contaminating the cultures with unwanted organisms. Single-celled and colonial forms are best transferred with a sterile pipette or dropper. Remove the pipettes or droppers from their storage chamber, making sure you do not contaminate the tips. Put the large end into a rubber bulb and draw water containing algae halfway up the glass tube. Do not draw water into the bulb. Hold the tip just under the surface of the new culture medium. Squeeze gently on the bulb to release the algae into the medium.

You can pick up filamentous algae with forceps or tweezers. Sterilize the tips of the forceps by dipping them first into a container of alcohol (baby food jar, for instance) then passing them through the flame of an alcohol burner. The alcohol will ignite briefly. Allow the forceps to cool a few seconds, then dip the tips into the water containing algae.

Pick up some strands of the algae and put them in the new culture. Another method of transferring is to use sterile cotton swabs from a drugstore; swirl the algal filaments around the cotton swab, then transfer the filaments to the new medium and unswirl.

Laurie Perkins

References

Growing Freshwater Algae in the Laboratory. Turtox Service Leaflet, No. 6. Turtox/Cambosco, Chicago, Illinois. 1961.

Hampton, Carolyn H., and Carol D. Hampton. "Collecting and Observing Algae." *Science and Children* 17:50-52; September 1979 (p.20 of this book).

James, Daniel E. *Culturing Algae.* Carolina Biological Supply Company, Burlington, North Carolina. 1974.

Morholt, Evelyn, et al. *A Sourcebook for the Biological Sciences.* Second Edition. Harcourt, Brace and World, New York City. 1966.

Aids for Identification

Hausman, Leon A. *Beginner's Guide to Freshwater Life.* G. P. Putnam's Sons, New York City. 1950.

Klots, Elsie B. *The New Field Book of Freshwater Life.* G. P. Putnam's Sons, New York City. 1966.

Reid, George K. *Pond Life.* Herbert S. Zim, Editor. (Golden Guide Series) Western Publishing Co., Inc., Racine, Wisconsin. 1967.

Duckweed

Duckweed is an easily grown, attractive aquarium plant that has many uses in elementary classrooms. There are four genera in Lemnaceae, the duckweed family. All are small to minute floating aquatic plants. One species, *wolffia* or water-meal, is about the size of a sand grain and is the smallest of the flowering plants. Other species range from two to four millimeters in size. The oval or elongated, flattened bodies of duckweed are made up of leaf-like stems called fronds. The fronds float on the water's surface while thread-like rootlets trail in the water.

The following key will help you determine duckweed's different generic groups:

Reprinted from *Science and Children*, Carol D. and Carolyn H. Hampton, April 1979, pp. 50-51.

Lemna is one of several prolific duckweeds.

Plants with roots and two reproductive pockets	
One root per frond	*Lemna*
Two or more roots per frond	*Spirodela*
Plants without roots and only one reproductive pocket	
Fronds globed or elliptical shaped	*Wolffia*
Fronds elongated and sickle shaped	*Wolffiella*

Duckweeds are widely distributed across the United States. The plants can be found on the surface of bodies of quiet fresh water—ponds, swamps, slow-moving streams, or at the margins of lakes. Duckweeds are eaten by ducks, geese, fish, and certain snails. In the country's colder regions, clusters of fronds sink to the bottom of bodies of water in the autumn. In the spring, the plants float to the surface and begin reproducing. Duckweeds can be collected by hand or dipped from the water with a kitchen strainer. Carry the plants to the classroom in a container of water from the collecting site.

Duckweeds grow well in freshwater aquaria under conditions of average classroom temperatures. Put the aquarium in a well-lighted area, but *not* in direct sunlight. Keeping the aquarium under a fluorescent ceiling light or in front of a north-facing window is best. The plants are decorative, cover to small fish, protect aquarium animals from too much light, and absorb some nitrogenous waste products produced by animals. Duckweeds do not add much oxygen to the water. Once established, duckweeds may reproduce so rapidly they become a nuisance. Skim unwanted plants from the water surface with a dip net.

Even though duckweeds are among the smallest flowering plants, reproduction by flowering is rare under classroom conditions. Propagation is usually by vegetative means. New

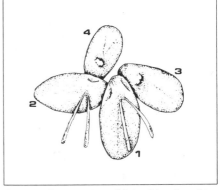

Figure 1. Bottom view of Spirodela oligorrhiza *(1 = oldest frond; 4 = youngest)*

fronds arise from growth regions—meristems—located in one or two pockets near the sides at one end of the parent frond. (See Figure 1.) As daughter fronds grow, they may remain attached to the parent frond for a short time before separating. This growth pattern causes duckweeds to grow in clusters. Surface tension may cause clusters to adhere, forming large green mats over the water.

Duckweeds' size, rapid growth, and ease of handling make them ideally suited for classroom use. Observe *Lemna* with microscopes to study plant cells, chloroplasts, stomata, and root

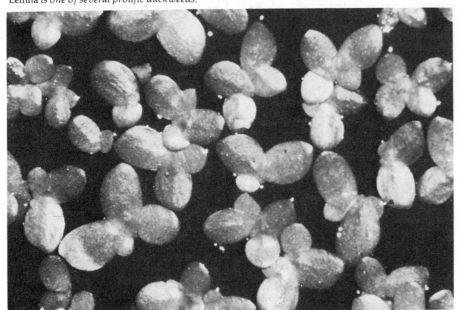

Courtesy Carolina Biological Supply Company

caps. Children can observe vegetative reproduction with a hand lens. Clones—offspring of a single parent—show "handedness." In a left-handed clone, a daughter frond will grow first from the left-hand pocket; then a daughter frond will grow from the right-hand pocket for each generation. (See Figure 1.)

Duckweeds are excellent for experimental population studies. Have children count individual fronds rather than trying to decide what makes up a separate plant. Students can count the fronds in their containers on the same day each week, recording data in a table similar to the one in Figure 2. If data are graphed, students will probably obtain a typical sigmoid (s-shaped) growth curve. Actual class data is shown plotted in Figure 3.

Keep the container with the duckweeds small, otherwise the number of fronds will become too numerous to count. If this happens, let children invent ways to estimate the population size. Students might try massing the total plant material. Use this data to indicate population growth. Baby food jars, small plastic vials, or small petri dishes (55 mm in diameter) make excellent growth chambers.

Given enough space, the plants will continue in an exponential growth pattern; duckweeds will double with every

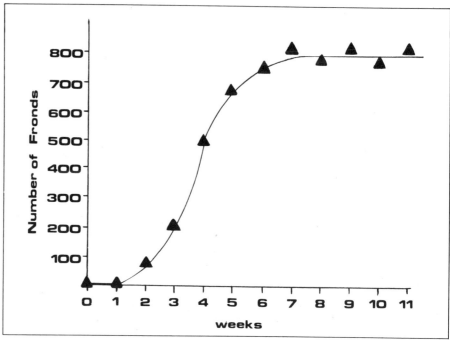

Figure 3. Population growth of duckweed.

generation. When the plants completely cover the surface of the water in a container, growth slows as the plants begin to compete for light and available minerals. As fronds get pushed on top of each other, those on the bottom die from insufficient light. The population size reaches a stationary level.

Upper middle grade and junior high school students could design a "super" medium or study the effect of plant nutrients and other environmental factors on duckweeds' population growth. A growth medium we have found successful for a 55 mm petri dish is made up of a pinch of sifted loamy garden soil, 3 pellets of 8-8-8 fertilizer (one of each color), and 15 mL of distilled water or aged tap water. Be sure children mark the water level in their growth chambers so they can replace water that evaporates. Always add aged tap water.

Curricula

Rockcastle, Verne N., et al. *Elementary School Science, STEM.* Addison Wesley Publishing Co., Inc., Reading, Massachusetts. 1975. Pp. 99-100.

Subarsky, Zachariah, et al. *Living Things in Field and Classroom.* Minnesota Mathematics and Science Teaching Project, University of Minnesota. 1969. P. 33.

Thier, Herbert D., et al. *SCIS: Populations,* Life Science Unit, Level 3. Rand McNally & Company, Chicago, Illinois. 1972. Pp. 66-67.

Thier, Herbert D., et al. *SCIS: Organisms,* Life Science Unit, Level 1. Rand McNally & Company, Chicago, Illinois. 1970. Pp. 29-32.

Resources

Behringer, Marjorie P. *Techniques and Materials in Biology.* McGraw-Hill Book Co., Inc., New York City. 1973. Pp. 322-323.

Morholt, Evelyn, et al. *Sourcebook for the Biological Sciences,* 2nd Ed. Harcourt Brace Jovanovich, Inc., New York City. 1966. P. 456.

Rhodes, Lee W. "The Duckweeds: Their Use in the High School Laboratory." *The American Biology Teacher* 30:548-551; May 1970.

Figure 2. Population size of duckweed.

week	Population Size Total Number of Fronds
1	
2	
3	
4	
5	
6	
7	
8	
9	
10	

Anacharis

Elodea (*Anacharis*) is an attractive plant well suited for classroom aquaria. It provides material for many elementary school experiments. Elodea grows submerged in aquariums, either rooted in the bottom substrate or floating on the surface. Its branches of dark whorled leaves provide a protective cover for fish and the fry (young) of aggressive adults.

The usefulness of plants to provide oxygen for animals in aquariums has been greatly exaggerated. Green plants, carrying on light-dependent photosynthetic reactions, release more oxygen than they use in respiration during the day. At night, the light phase of photosynthesis stops. Plants continue to use oxygen in respiration and may actually compete with animals for the available oxygen supply. Keep in mind that unless an aquarium receives plenty of light, the green plants may be a detriment.

Benefits that plants are apt to provide in an aquarium are the taking up of carbon dioxide and nitrogenous wastes produced by the animals. Increased amounts of carbon dioxide dissolved in water mean there is less oxygen—and lowered oxygen content and increased nitrogenous products may cause death of the animals.

Spacing elodea in aquariums may depend on your specific objective. If elodea is to be used primarily for esthetic purposes, then push the cut ends of the sprigs into the sand or gravel. Leave at least 10 cm between plants. For a more attractive, less cluttered appearance, plant the long sprigs near the back and the small ones near the front. If you need to provide cover for young or shy fish, leave some sprigs floating at the surface. Floating plants may also shield fish from excess light. Roots will form on the ends placed in the sand. Roots will grow downward from various regions along the stems of floating plants.

Collection

Elodea may be purchased from aquarium shops or biological supply houses. It can be found growing naturally along the edges of lakes and ponds and in the shallows of sluggish streams. Under ideal conditions, it may produce flowers that float on the surface attached to the stems by long slender stalks. However, propagation is primarily by vegetative means, by branching.

Collection usually requires wading. If wading boots are unavailable, use old tennis shoes. Pinch 20-cm sprigs off the greenest, healthiest looking branches. Submerge them in a plastic bucket containing water from the collecting site. You can also wrap clusters of sprigs in damp paper towels which are then wrapped in newspapers to carry back to the classroom.

Before using plants from a natural source in an aquarium, wash the sprigs thoroughly by running tap water over them until mud and algae are removed. If the elodea is badly contaminated with algae, rinse the sprigs in a weak solution of potassium permanganate for the final rinse (one grain per 3.8 liters of tap water). If potassium permanganate is unavailable, use a commercial product from an aquarium shop. Follow the directions on the label.

Maintenance

Several elementary science curricula require use of elodea in sufficient quantities for teams of students to keep their own aquaria. In some cases, elodea may be needed by several different grade levels and by a number of schools in the same educational unit.

Elodea may be cultured in quantity. Place 80- to 95-liter aquarium tanks or large battery jars in a well-lighted room—preferably under fluorescent lights. Layer the bottom of the tanks with 1 cm of rich garden soil or potting soil.

Place a 1.5 cm layer of pebbles over this. Next, add a 2.5 cm layer of sand. To retard growth of single-celled and colonial forms of algae, place a strip of copper (about 2.5 cm × 5 cm) just below the surface of the sand.

Be sure to wash the pebbles and sand thoroughly. A good way is to put the sand or pebbles in a small plastic bucket. Run warm water over the material. Let the sand or pebbles settle to the bottom. Pour off the water. Continue washing until the water remains clear.

The aquaria should be filled with dechlorinated water. If spring or pond water is unavailable, use tap water. Let tap water stand in uncovered plastic buckets for three days so the chlorine escapes. Cap the buckets with covers to prevent evaporation and contamination until you are ready to set up the aquaria in the classroom.

Place a shallow saucer on the sand layer. Pour in the dechlorinated or spring water until water is about 15 cm deep. Plant the sprigs of elodea by pushing the cut ends into the sand. Be sure not to disturb the substrate layers. Space plants about 8-10 cm apart. Add water until the level is about 1.5 cm from the top. Cover the tank with a piece of glass.

This way the soil will provide nutrients for plant growth. Be sure to keep the tanks well lighted. Poorly lit elodea tends to become long and spindly.

The plants should elongate rapidly and produce roots within two weeks. To harvest, pinch off 15-20-cm sprigs from the tops of the rooted plants, without disturbing the substrate layers. Remove any filamentous algae by hand as soon as it can be seen accumulating.

The tanks should be kept at room temperature, ideally between 20-21°C. If the water becomes too warm, the plants may get spindly, lighten in color, and decay.

Reprinted from *Science and Children*, Carol D. and Carolyn H. Hampton, November/December 1978, p. 27.

Seed Plants

Plants are inexpensive to keep and are convenient subjects for student investigations. Students may study firsthand plant structures, plant functions (e.g., photosynthesis, transpiration, seed respiration), methods of reproduction, vegetative propagation, and the effects of various environmental conditions on plant growth and development.

Materials
- Potting soil (sterile)
- Coarse building sand
- Vermiculite
- Three plastic trash cans with lids
- Three soup cans (identical size)
- Several plastic containers (buckets, dishpans, large bleach bottles with tops cut off)
- Plant containers (clay and plastic pots, paper and styrofoam cups, milk cartons)
- Trowels, large plastic spoons or scoops
- Foam plastic trays from meat market
- Newspaper

If you are enthusiastic about working with plants in the classroom and are going to involve your pupils in making cuttings and planting seeds, then a good way to keep the materials for planting mixes is to store them in plastic trash cans with tight fitting lids. (See Figure 1.) You will need to store potting soil, sand, and vermiculite in separate cans. Keep a soup can of equal size in each trash can to serve as a scoop and a measuring device. Label the containers and store them under a workbench or table. A long table in front of a set of windows provides a surface for

| | Planting Mixes | |
Seeding Mix	Rooting and Cutting Mix	Potting Mix
2 cans sterile potting soil 1 can sand 1 can vermiculite	1 can sand 1 can vermiculite	3 cans sterile potting soil 1 can sand 1 can vermiculite

the growing plants and a space under which the planting materials may be stored. Store plant containers in a large cardboard box and smaller items in shoe boxes.

The methods described in this article may vary from those described in greenhouse manuals and gardening books. Lack of methods for sterilizing soil, time, and expense are important factors that must be considered in planning for school classroom activities. If a greenhouse and a method of sterilizing are available, consult professional plant care manuals.

Use a sterile potting soil (check label on bag), vermiculite, and a coarse building sand for making up the planting mixes. Place canfuls of the materials in plastic buckets or dishpans for children to mix thoroughly using a scoop or large spoon.

Germinating Seeds

Plastic shoe or sweater boxes and ziplock bags make excellent germination chambers. Place several layers of paper towels in the bottom of the container. Dampen the paper thoroughly and place the dry seeds on the paper. Space larger seeds about 3 cm apart; smaller seeds about 2 cm apart. Repeat the layers of dampened paper and seeds until you have enough seeds for the intended investigation. Cover the box or seal the bag and place the germination cham-

Figure 1. Plants and Materials Storage.

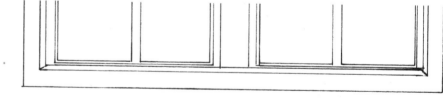

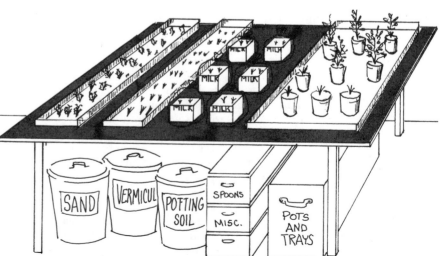

Reprinted from *Science and Children*, Carol D. and Carolyn H. Hampton, March 1981, pp. 25-27.

bers in a dark place. Most seeds will germinate within 48 hours. For individual or small group germination chambers, use petri dishes or small plastic food containers. Seeds may also be germinated on a layer of dampened vermiculite.

Planting Seeds

For individual plant containers, use paper cups or milk cartons; for larger numbers of plants in a confined space, use seed flats. Inexpensive plastic flats may be purchased at garden shops or constructed from cardboard milk cartons with one side removed. Be sure to puncture the bottom with several holes for drainage. Add 6 to 10 cm of seeding mixture to the containers. Place a seed flat in a pan of water until the soil becomes damp throughout. With a pencil, mark off planting rows about 5 cm apart. Place the seeds in rows about 1 to 2 cm apart. Cover the seeds with soil to the depth of twice the thickness of the seeds. For large seeds and seeds with hard seed coats such as corn, beans, and squash, soak them for 12 hours (overnight) before planting. For very fine seeds such as grass, radish, leaf lettuce, mustard, and petunias, sprinkle the seeds over the soil and press them into the soil gently with a block of wood.

To provide a humid atmosphere, insert the flats into a plastic laundry bag and tie off the opening. Place the bag in darkness or diffuse light and inspect the flats daily. If necessary, moisture can be added with a pump spray bottle. Excessive moisture will promote the growth of fungi and may even cause the seeds to decay. If moisture condenses on the sides of the plastic, leave the bag open for several hours.

Excessive moisture, crowding of plants, and the use of unsterile pots and planting trays promote the fungal disease called "damping off." The disease causes the stems of the young plants to break off at soil level and die.

When half the seeds have sprouted, remove the trays from the plastic bag and place them near a window in indirect sunlight or under fluorescent lights. Allow the top 1 cm of the soil to dry before watering so that the roots will grow downward. Remove weak plants and plants from crowded areas. Allow about 5 cm of space between individual plants.

Certain flowering annuals may be grown from seeds in the classroom. If started in late autumn or early winter, they will flower in two to three months. If given plenty of sunlight, the following plants will grow well at room temperatures: dwarf marigolds, nasturtiums, geraniums, petunias, morning glories, sweet alyssum, calendula, dwarf corn flowers, and moss rose. The sensitive mimosa grows well from seed and provides an excellent opportunity for children to observe rapid movement in plants in response to touch (thigmotropism).

For seeds that are fun to grow and make attractive foliage plants, try fruit plants such as kiwi, calamondin orange, lime, Meyer lemon, papaya, and date. Place potting mix in a container and place the seed (or several if the seeds are small) on the soil while still damp from the ripe fruit. Push the seeds into the soil to a depth twice their thickness and cover. Before planting a date seed, use a sharp blade to scrape away some of the hard seed coat to allow moisture to enter the seed. Remember, all of these plants are tropical and will need warm temperatures to grow. Another interesting tropical seed is the avocado. Insert four toothpicks in the large seed and place these on the rim of a plastic cup filled with water so that the pointed end of the seed rests in the water.

Transplanting Seedlings

For short-term studies, young plants grown in flats do not need to be transplanted. When mature plants are required, transfer the seedlings to individual containers between the time that the first and second foliage leaves appear and the plants are between 5 to 8 cm high. If the transfers are made later, the matted roots will be damaged when removed from the soil. Plastic or clay pots are best for mature plants; however, small milk cartons with holes punched in the bottom will suffice.

Place a large pebble or a piece of broken pot over the hole in the bottom of a small pot to provide drainage and retain the soil. For a larger pot, place a layer of pebbles in the bottom. Add the potting mix to about two-thirds of the depth of the pot and shape the soil up on the sides leaving a large hole in the center. Remove the young plant from the tray with a spoon and place it in the new container at the same depth that it was in the previous container. Add soil to fill in around the plant, pressing gently until the level is about 2 cm below the rim of the pot. Water the soil and place the pot in diffuse light for the first week. Then keep the plant in the light conditions recommended for the species. At the end of the first week, water the plant with liquid plant food. Continue to use the liquid plant food every three to four weeks or according to the directions on the package.

Plants that are especially easy to care for in classrooms are coleus, begonia, geranium, tradescantia (wandering Jew), impatiens, and spider plants. For rooms with limited sunlight and varied temperatures, philodendron, arrowhead vine, snake plant, small types of ivy (English, German, Boston), and Chinese evergreen are recommended.

Plants should be inspected daily with a magnifying glass for insect pests. Remove any insects by hand. Mealybugs appear as small cottony masses and should be wiped off with a cotton pad dampened with medicinal alcohol. Otherwise, a weekly cleansing using a soft cloth dampened with lukewarm water will keep the foliage free from insects and dust.

Vegetative Propagation

Most methods of vegetative propagation require high humidity. Prepare propagation chambers by cutting off the top two-thirds of a cardboard milk carton and puncturing the bottom with several holes. Fill the carton with rooting and cutting mix and add water. Turn the mix over several times with a spoon to insure that it is thoroughly dampened, but there is no standing water in the container. Place from

Figure 2. Propagation Chamber.

three to four cuttings about 2.5 cm deep and place the carton into a plastic bag. Inflate the bag by blowing into it and fasten the open end with a twist tie or rubber band. (See Figure 2.)

Stem Cuttings

Take cuttings from well developed side branches of plants. With a sharp, clean blade, cut off stem tips that have at least two leaf nodes, about 7 to 12 cm in length. Carefully tear off the lower leaves and cut off any stem below the lowest node.

Push herbaceous stems (coleus, geranium, begonia, impatiens, tradescantia) cuttings immediately into the sand and vermiculite mixture. Most should produce roots within one to two weeks. Dip softwood stems (ivy, privet, willow, golden bell) into a plastic cup containing a small amount of rooting powder (hormone). Tap the stems with a pencil to remove the excess powder and then push the stems into the rooting and cutting mixture against the bottom of the container. Press the soil firmly about the stems. Pressure stimulates a callus to form over the cut surface of a stem. After three weeks, examine a stem to see if the callus has formed. If so, roots will probably develop in another two to three weeks.

Leaf Cuttings

Plants with fleshy leaves and stems, including African violets, begonias, *Kalanchoe*, *Sedum*, and practically all other succulents, may be propagated by leaf cuttings. Select a fully developed leaf and cut through the stem about 2 to 3 cm below the leaf. Insert the stem and the base of the leaf into the rooting and cutting mix. New plants should appear at the base of the leaf in several weeks. When the plantlets have grown to about one-half the size of the original leaf, separate and transplant them.

Place the propagation chambers for stem and leaf cuttings in indirect light until roots or plantlets form. Move the chambers to more direct light and open the bags for half a day on the first day. On the next two days, leave the bags open to allow the plants to adjust to drier conditions. After this, transplant the cuttings or plantlets to containers of potting mix using the same procedure described under transplanting.

Other Plant Studies

Narcissus, hyacinth, and Easter lily bulbs may be grown in the cutting/

Figure 3. Plant parts suspended in water.

rooting mix. Cover the bulbs with 5 cm of soil in pots or milk cartons. Keep the bulbs of hyacinth and Easter lilies in a cool dark place for about four weeks and then transfer them to a well-lighted, warm location. Flowers should appear in three to four months. Narcissus grows quickly and should flower within a month in a warm room.

Students can plant white potatoes in milk cartons containing the potting mix. Cut a potato into pieces, making sure that each piece has one to several "eyes." Arrange each piece so that the eyes are pointed upward and cover the piece with 3 cm of soil.

Sweet potatoes, carrots, and radishes will develop new shoots if suspended in a container of water by four toothpicks. (See Figure 3.)

Bibliography

Behringer, Marjorie P. *Techniques and Materials in Biology.* McGraw-Hill, Inc., New York City. 1973.

Morholt, Evelyn, et al. *Teaching High School Science: A Sourcebook for the Biological Sciences.* Second edition. Harcourt Brace Jovanovich, Inc., New York City. 1966.

Subarsky, Zachariah, et. al. *Living Things in Field and Classroom.* Minnesota Mathematics and Science Teaching Project, University of Minnesota, Minneapolis. 1969.

Curriculum Aids

Budding Twigs. Elementary Science Study. Webster Division, McGraw-Hill, Inc., New York City. 1970.

Kramer, David C., "The Coleus: A Useful Plant for Classroom Activities." *Science and Children* 15:29-30; April 1978.

Nickelsburg, Janet. *Nature Activities for Early Childhood.* Addison-Wesley Publishing Company, Reading, MA. 1976.

Oaks, Acorns, Climate and Squirrels. National Wildlife Federation, Washington, DC. 1971.

Plant Puzzles. National Wildlife Federation, Washington, DC. 1972.

Plants in the Classroom. National Wildlife Federation, Washington, DC. 1971.

Simmons, Barbara, and June Hogue. "Motivating Green Thumbs." *Science and Children* 15:7-11; January 1978.

Collecting and Observing Protozoa

Ever since Leeuwenhoek first observed tiny "living creatures" in a drop of water and described them to the Royal Society of London, students have been fascinated by protozoa. Most protozoa are microscopic in size, but a few are large enough to see with the unaided eye. To observe and appreciate the beauty and complexity of these organisms, you need a microscope. An inexpensive student microscope equipped with low and high power objectives provides enough magnification for elementary pupils.

Finding Protozoa

Protozoa live almost everywhere there is moisture—in ditches, temporary pools, ponds, lakes, salt water, damp soil, mud, hot springs, snow drifts, glaciers, brine pools, and sluggish streams. Water is necessary for most of them during their active and reproductive stages. The thin film of water surrounding soil particles contains many protozoa. However, you rarely find them below the top 2.5 cm of soil.

Most vegetation debris collected from outdoor habitats will contain some resting stages or spores of protozoa. The best collecting sources are natural places—ditches, shallow ponds, and temporary pools, for example—containing plant remains in various stages of decomposition. Stagnant farmyard puddles contaminated with animal wastes are rich sources of protozoa. Some protozoa are parasites of other organisms.

Flagellates and ciliates are found in open surface water, on or among floating and submerged vegetation, on rocks and other bottom debris. You can find amoeboid forms on bottom sediments and debris and on submerged vegetation, including the undersides of leaves. You can find the stalked ciliates attached to dead sticks and submerged material, on the undersides of duckweed and floating plants, and on bodies of water insects, snails, and crustaceans. During the winter protozoa may still be found on debris and bottom sediments of bodies of water.

Small jars with tight-fitting lids, such as baby food jars, are suitable for collecting water samples of protozoa. Plastic buckets with lids make useful containers for carrying large samples and for collecting water from the habitat for classroom use. Do not fill containers of water samples more than half full. Seal tightly only when transporting samples back to the classroom. Open the containers as soon as possible after returning to the classroom to permit aeration. You can keep aquatic collections of protozoa near a window, but not in direct sunlight where they may become overheated. A northern exposure is best.

You can collect floating vegetation by gently lowering a container below the water's surface and letting the material flow into the container. A jar lid makes a handy scoop for gathering surface scum

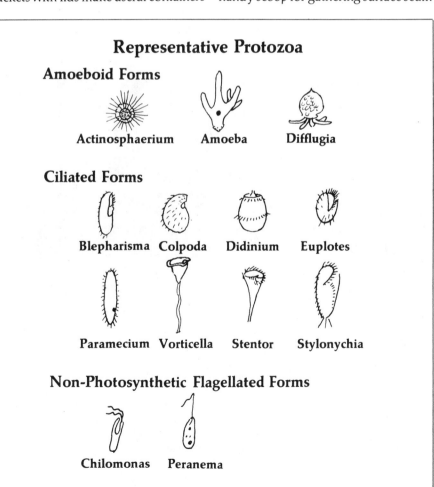

Representative Protozoa

Amoeboid Forms

Actinosphaerium Amoeba Difflugia

Ciliated Forms

Blepharisma Colpoda Didinium Euplotes

Paramecium Vorticella Stentor Stylonychia

Non-Photosynthetic Flagellated Forms

Chilomonas Peranema

Reprinted from *Science and Children*, Carol D. and Carolyn H. Hampton, November/December 1979, pp. 30-32.

and floating material. A small fine-mesh dip net is useful for scooping up material from pools and around the water's edge. To collect protozoa from open surface water, use a No. 20 mesh plankton net.

Bottom sediments may be gathered by scraping the surface with a small can attached to a long pole. A baster or glass tube 30 to 40 cm long attached to a soft rubber bulb makes a useful device for collecting on and above soft substrates. Squeeze the air out of the bulb. Slowly draw the sample into the tube by relaxing your grip on the bulb as you move the tube over the bottom. Transfer the contents to a collecting jar. You can scrape samples from rocks or other submerged objects with a knife or spatula.

Dry culture materials such as leaves, grass, twigs, rocks, and soil may be collected and put in any type of container—can, jar, or a paper or plastic bag. Start cultures of protozoa from dry collections by putting the materials in jars or flat enamelware pans and covering them with pond, lake, spring, rain, or aged tap water. Let the flooded material stand undisturbed in a warm place out of direct sunlight for 24 hours before examination.

Examining protozoa requires thoroughly clean glass slides, cover-glasses, and microscopes. When viewing large specimens, add small pieces of broken coverglass around the edge of the culture drop to prevent the regular coverglass from crushing them.

Compare protozoa from different soil types and aquatic habitats. Have students make wet mounts and hanging drop slides to study cultures. Make sure samples for observation are taken from the top, middle, and bottom of the cultures in order to see which kinds of protozoa inhabit those areas. Examine scrapings and pieces of vegetation for attached organisms. Look at each source regularly so students can observe changes in population counts, numbers

Using a plankton net. *Elizabeth Lawrence*

of species, and patterns of succession.

Slow down rapidly swimming protozoa for closer study by adding cotton or lens paper fibers to the slide before lowering the coverglass. An excellent way to slow down protozoa for detailed inspection is to mix a drop of methyl cellulose with a drop of culture before observing.

An interesting and simple way to demonstrate the organelles, or cell structures, in a living protozoan is to use vital stains available from scientific supply companies. Vital stains do not kill an

organism immediately and are taken up by different parts of the protozoan. Use methylene blue to demonstrate the nucleus, cytoplasmic granules, and cilia. Neutral red stains the food vacuoles making for easy identification. Methylene green stains the cytoplasm a light green. Add the stain to the drop of culture being observed before you add the coverglass. You also can add the stain at the edge of the coverglass and draw it under using a piece of absorbent paper toweling on the opposite side.

Many techniques used in collecting

and observing algae apply to protozoa. The September 1979 issue of *S&C* contained a companion article in the Care and Maintenance series, "Collecting and Observing Algae" (p. 20 in this book). The article gives details on building a field collecting tray and a plankton net, and how to prepare deep well and hanging drop slides for observing motile microscopic organisms.

Resources

Behringer, Marjorie P. *Techniques and Materials in Biology.* McGraw-Hill, Inc., New York City. 1973.

Curtis, Helena. *Marvelous Animals: An Introduction to the Protozoa.* Natural History Press, Garden City, New York. 1968.

Hall, Richard P. *Protozoa: The Simplest of All Animals.* (Holt Library of Science). Holt, Rinehart & Winston, New York City. 1964.

Kudo, Richard R. *Protozoology.* Fifth Edition. Charles C Thomas, Publisher, Springfield, Illinois. 1966.

Morrison, Sean. *Amoeba: A Photomicrographic Book.* Coward, McCann & Geoghegan, Inc., New York City. 1971.

Orlans, Barbara. *Animal Care: From Protozoa to Mammals.* Addison-Wesley Publishing Co., Inc., Reading, Massachusetts. 1977.

Pennak, R. W. *Fresh-Water Invertebrates of the United States.* John Wiley & Sons, Inc., New York City. 1953.

Sleigh, M. *The Biology of Protozoa.* Elsevier North-Holland, Inc., New York. 1973.

Identification Aids

Hausman, Leon A. *Beginner's Guide to Freshwater Life.* (Putnam's Beginner's Guide to Nature Series) G.P. Putnam's Sons, New York City. 1960.

Jahn, Theodore L., and Frances F. Jahn. *How to Know the Protozoa.* (Pictured Key Nature Series) William C. Brown Company, Publishers, Dubuque, Iowa. 1949.

Needham, James G., and Paul R. Needham. *Guide to the Study of Freshwater Biology.* Fifth Edition. Holden-Day, Inc., San Francisco, California. 1962.

Reid, George K. *Pond Life.* Herbert S. Zim, Editor. (Golden Guide Series) Western Publishing Co., Inc., Racine, Wisconsin. 1967.

Curriculum Ideas

Goldstein, Philip, and Jerome Metzner. *Experiments with Microscopic Animals.* Doubleday & Co., Inc., New York City. 1971.

Pond Water. (Elementary Science Study Unit) McGraw-Hill, Inc. New York City. 1967.

Culturing Protozoa

ield trips to ditches, ponds, and streams during spring and fall provide excellent opportunities for collecting protozoa. Such collections can serve as sources of organisms for culturing. Important factors to consider when culturing protozoa are light, temperature, pH of the culture medium, cleanliness and type of containers, water source, food supply, and contamination by other organisms.

Requirements

Light. Generally, most cultures should be placed in medium light areas away from direct sunlight. Some protozoa grow quite well in darkness *(Amoeba)* while chlorophyll-bearing forms *(Euglena)* do well in light from a window facing north. A fluorescent lamp works well in the absence of adequate natural light. Cultures kept in strong light are likely to develop growths of unwanted algae. Do not let cultures overheat and dry out.

Cupboards or shelves with doors make good places to store cultures requiring little or no light. Cover the surfaces where the cultures will be stored with "pest control" paper to keep ants from invading the containers.

Temperature. Most protozoa may be cultured satisfactorily at an optimum range of room temperatures between 18° and 22 °C. If the temperature drops below 15 °C, the cultures will stop multi-plying and eventually die. At temperatures above 30°, the cultures will die rapidly.

pH. The culture medium for most protozoa should be neutral or slightly basic (pH 7.0 to 7.6). Check the pH with hydrion paper.[1] If the medium is too acidic, correct it by adding a solution of water and sodium hydroxide mixed in the proportion of 4 g NaOH to 100 mL of water (1N NaOH). Add one drop of solution at a time.

Culture Vessels. Any glass containers such as baby food jars or peanut butter jars, with loose fitting lids, may be used for culturing protozoa. If they are available, culturing dishes (12 × 5 cm) stacked on top of one another make ideal containers. The containers, instruments, and workspace should be clean and uncontaminated.[2]

Make a Micropipette

A micropipette is convenient for transferring small protozoans. (See Figure 1.) One can be made by heating the center region of a 20-cm long piece of 4 mm (outside diameter) glass tubing until it begins to melt. Quickly pull both ends of the tubing with a steady pressure stretching the center and you will have two micropipettes when the glass breaks. The large ends should be fire polished to prevent cutting your hands when handling them.

Insert the large end of the micropipette into a piece of soft plastic aquarium tubing attached to a syringe. Practice using the instrument before transferring cultures. By using the micropipette with a dissecting microscope, you can capture and transfer single organisms.

Water. Do not use tap water for culturing, as it contains chlorine and other chemicals which may kill the organisms. Natural sources of water such as pond, lake, spring, or rainwater may be used after they have been filtered, pasteurized (heated to 73 °C for 20 minutes) and cooled to room temperature. In some cases, use distilled water for special media. Check the specific formulas for media to see if the water should be made aseptic before or after mixing with other ingredients. After heating the water, store it in containers covered with loose fitting lids, or glass plates, which will allow gas exchange while keeping out contaminants. The water level in the culturing vessel should be marked on the outside. To compen-

[1] Available from Carolina Biological Supply Company, Burlington, North Carolina, Ward's Natural Scientific Establishment, Rochester, New York, or other supply companies.
[2] See "Care and Maintenance," *Science and Children*, February 1980 (p. 23 of this book), for complete instructions on preparing glassware and the workspace before attempting the culturing of protozoa.

Figure 1. Completed Micropipette.

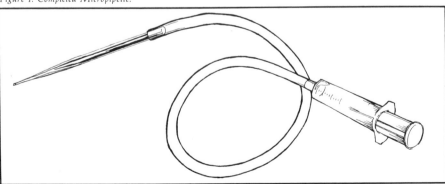

Reprinted from *Science and Children*, Carol D. and Carolyn H. Hampton, April 1980, pp. 34-36.

ate for evaporation, aseptic or distilled water may be added when needed.

Culture Media

The following media have been successful in culturing certain protozoa under regular classroom conditions.

Sterile Wheat, Rice, and Hay. Cover the bottom of a baby food jar with wheat seeds bought from a health food store. Add enough distilled or deionized water to cover the seeds. Place the lid on the jar but do not tighten it. The jar may then be placed in a pressure cooker and sterilized by maintaining the pressure at 15 pounds per square inch at 121 °C for

Figure 2. Heating Tubing in Center.

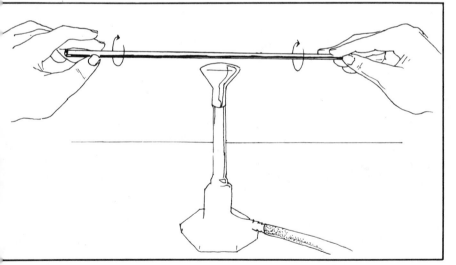

Figure 3. Pulling Tubing to Pipette Size.

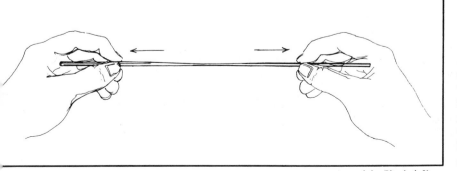

15 minutes. After sterilizing, remove the jar and tighten the lid. Allow the jar to cool. Use the same procedure for sterilizing rice grains and timothy hay clippings. Jars containing sterilized wheat, rice, and hay may be stored on a shelf until needed.

When you are ready to transfer sterile seeds and hay from their containers, we recommend the following method:
1. Dip forcep tips into a baby food jar of alcohol.
2. Pass the tips through the flame of an alcohol burner.
3. Allow the alcohol to burn from the tips.

4. Remove the lid from the sterilized jar of seeds or hay. Pick up the seeds and place them into the culture vessel.
5. Replace the lid on the jar and tighten.

As long as no fuzzy growth appears on the seeds and hay, you may assume they are sterile and can continue using them until they are gone or become contaminated.

Wheat Medium. Fill the culture vessel half-full with sterilized spring or rain water. Add three or four sterile wheat seeds. Inoculate the medium with the desired organism using a sterile dropper or pipette. Introduce at least 50 organisms of the same kind into the new medium in order to establish a healthy growing culture. Cover the vessel, place it in an appropriate place, and allow it to remain for one or two weeks before examining. Use for *Actinospharium, Amoeba, Blepharisma, Chilomonas, Colpidium, Colpoda, Euplotes, Peranema, Stentor,* and *Vorticella.*

Wheat-Hay Medium. Follow the same general procedure as for preparing a regular wheat medium. Add three or four sterile hay sprigs along with the wheat seeds. If culturing *Difflugia*, add 1 g of clean sand to the bottom of the culture dish for the organism to use in constructing its shell or "test." Inoculate the medium as instructed before and store the culture in an appropriate place. Use for *Arcella, Blepharisma, Difflugia, Paramecium,* and *Spirostomum.*

Grain-Dried Milk Medium. Add 40 wheat seeds, 40 rice grains, and 1 g of dry nonfat powdered milk to 1 L of distilled water. Stir until the milk is dissolved. Boil this mixture for five minutes, cover, and let cool to room temperature. Inoculate with the organisms to be cultured. This medium works well for *Euglena* and the culture should be placed in a well-lit area but not in direct sunlight. Use for *Chilomonas,*

Artwork by Elizabeth Kays.

Colpidium, Colpoda, Euglena, and *Paramecium.*

Soil-Water Medium. This medium is described in "Growing Algae in the Classroom," *Science and Children,* February 1980. The medium may be considered an artificial pond and is suitable for culturing many protozoa and algae. Use for most ameboid and ciliated protozoans plus the algae *Chlamydomonas* and *Volvox.*

Chalkley's Medium (Synthetic Pond Water). This medium is widely used for culturing *Amoeba.* Where sophisticated balances are unavailable, the authors recommend that a concentrated stock solution (10× or 100×) be prepared and then diluted to culturing strength before using. (See Chart.)

To make culturing strength solution, add 100 mL of the 10× solution to 1 L of distilled water (or add 10 mL of the 100× solution to 1 L sterile distilled water.) Add either sterile wheat or rice grains (4 seeds to 200 mL of solution) to the medium before adding the organism to be cultured. Use this solution for *Amoeba* and place in the dark. (Use for *Chilomonas* and *Colpidium* and place in medium light as food for *Amoeba.)*

Additional examples of culture media may be found in the references. The preceding techniques and media have proved satisfactory in the author's

Dilution Chart for Chalkley's Medium

Regular Strength Formula	For 10× Stock Solution (Multiply by 10)	For 100× Stock Solution (Multiply by 100)
0.1 g NaCl (Sodium Chloride)	1 g NaCl	10 g NaCl
4 mg KCl (Potassium Chloride)	40 mg KCl	400 mg KCl
6 mg CaCl$_2$ (Calcium Chloride)	60 mg CaCl$_2$	600 mg CaCl$_2$
1 L of dist./sterile H$_2$O	1 L of dist./sterile H$_2$O	1 L of dist./sterile H$_2$O

classes over the years. By maintaining cultures on a routine basis, living protozoa will be available for class studies throughout the school year.

Transferring cultures to new containers of media should be done every two to six weeks depending on the growth rate of the individual organism. Use sterile containers and instruments to avoid contamination of new cultures from unwanted organisms. None of the cultures will remain "pure" for a long time; however, the desired organism will usually be the predominant form and can be studied and subcultured at regular intervals.

References

1. Behringer, Marjorie P. *Techniques and Materials in Biology.* McGraw-Hill, Inc., New York City. 1973.
2. *Culture of Protozoa in the Classroom.* Culture Leaflet No. 1. Ward's Natural Science Establishment, Inc., Rochester, New York.
3. *Culture Media for Protozoa and Algae.* Culture Leaflet No. 5. Ward's Natural Science Establishment, Inc., Rochester, New York.
4. Hampton, Carolyn H., and Carol D. Hampton. "Growing Algae in the Classroom." *Science and Children* 17:40-41; February 1980 (p.23 of this book).
5. Morholt, Evelyn, et al. *A Sourcebook for the Biological Sciences.* Second Edition. Harcourt Brace Jovanovich, Inc., New York City. 1966.
6. Orlans, Barbara. *Animal Care: From Protozoa to Mammals.* Addison-Wesley Publishing Company, Inc., Reading, Massachusetts. 1977.
7. Schwab, Joseph J. *Biology Teacher's Handbook.* BSCS. Third Edition. John Wiley & Sons, Inc., New York City. 1978.
8. *The Care of Protozoan Cultures in the Laboratory.* Turtox Service Leaflet No. 4. Macmillan Science Company, Chicago, Illinois. 1959.
9. Whitten, Richard H., and William R. Pendergrass. *Care of Lower Invertebrates.* Carolina Biological Supply Company, Burlington, North Carolina. 1977.

Planaria

Children are fascinated by the "cross-eyed" flat worm commonly called planaria. With hand lenses, they may observe the worms' anatomy, feeding behavior, and reactions to environmental stimuli.

Planarians may be found attached to submerged vegetation, the underside of twigs, dead leaves, and stones, or crawling on the bottom mud at the edges of ponds, lakes, and springs. The most common one and the easiest to keep in the classroom is the brown planarian *(Dugesia tigrina).* The brown planarian has a triangle-shaped head, two dark light-sensitive spots called eyes, and prominent ear-like projections called auricles.

(See illustration.) Their color may be yellowish-brown, olive gray, or dark brown. They are about 1.5 centimeters in length.

Collection

Try to collect the worms by dangling a piece of raw beef liver, lean red meat, or hard-boiled egg yolk (in a cheesecloth bag) on the end of a string into a likely pond. If planarians are there, they should collect on the bait and can then be washed off into a plastic container filled with water from the collecting site.

Our preferred collection technique is to gather submerged vegetation and debris from the edge of the pond and put it in a plastic bucket holding water from the site. Later, in the classroom, we pour the bucket contents into plastic dishpans. We cover the containers with glass plates or cardboard, and let them sit for 24-48 hours. The oxygen content of the water eventually lowers and the worms glide to the surface.

Maintenance

To keep planarians in the classroom for some time, you can culture them successfully in shallow pans or bowls

Reprinted from *Science and Children*, Carol D. and Carolyn H. Hampton, October 1978, pp. 39-40.

containing spring, pond, or aged tap water (water that has remained in an open plastic bucket for at least three days to allow chlorine gas to escape). Culture vessels may be stackable glass culture dishes (200 mm × 85 mm) or glass or enamel baking pans.

Fill the containers 3 cm deep with water. A large surface to volume ratio permits adequate exchange of oxygen and carbon dioxide. Place the dishes in the darkest area of the room. If the vessels are glass and the room is well lighted, add one or two large dead oak leaves, large chips of wood, or pieces of broken flower pots to provide places for the planarians to attach themselves. They are negatively phototaxic—they tend to move away from the source of light and will attach to the underside of opaque objects. Add about 25 to 50 worms to each dish or pan. The worms thrive best at a room temperature of 20° C to 22° C (70°–72°).

Once a week, add a piece of raw beef liver, red lean meat, or hardboiled egg yolk about the size of a pea. The planarians swim to the food source and will collect on it while feeding. After two or three hours, remove the food and change the water.

Any objects such as leaves or cover for the planarians may be temporarily removed to another container while changing the water. Run a finger around the bottom of the container to loosen any slime that may have collected. Pour out the used water and rinse the culture dish with fresh water. Be careful not to dislodge the planarians attached to the sides. If you pour both the old and rinse water into a shallow pan, any dislodged worms can be recovered.

Transfers may be assisted with a wide-mouth dropper, but the worms have a tendency to adhere to the sides. The easiest technique is to pick up the animals with a long fingernail or the broad end of a flat toothpick.

For student observations or regeneration experiments, planarians may be slowed down to help handling by placing them on ice cubes or in the refrigera-

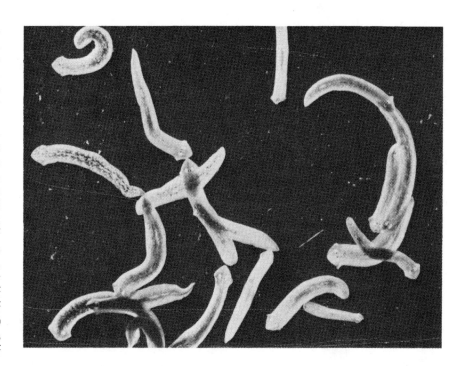

tor. Several grains of Epsom salts added to the water in a small dish containing the worms acts as a narcotic. You can make their digestive tracts more visible by feeding them for several weeks on an egg yolk diet before observations.

Suggested Activities

1. Design a *trap or bait* to catch planarians (either from an outdoor pond or a classroom "minipond").
2. Design a *habitat* for the worms.
3. Conduct library research followed by an oral or written report.
4. Conduct regeneration experiments. Cut large healthy worms into three sections (head, middle, tail) by making transverse cuts with a razor blade. Place each section into different dishes with water. See what happens to each section. Be as clean as possible with this procedure. Change the water often to avoid contamination and to help healing. Withhold food during the regeneration process. Remove any dead animals (gray or fuzzy).
5. Observe worms' behavior in response to:
   a. Touch.
   b. Food. Try feeding them different foods: beef, liver, hard-boilded egg yolk, daphnia, brine shrimp, pieces of earthworm, oatmeal, others.
   c. Light and dark.
   d. Weak electrical stimulation.
   e. A small piece of absorbent paper soaked in vinegar or household ammonia.

For further reading on activities students can do with planaria:

1. Brennan, Matthew, and Paul F. Brandwein. *A Searchbook, Life, Its Forms and Changes.* (Student Manual). Harcourt, Brace, and World, Inc., New York. 1969.
2. Minnesota Environmental Sciences Foundation, Inc. *Change in a Small Ecosystem.* National Wildlife Federation, 1972.
3. Subarsky, Zachariah et al. *Living Things in Field and Classroom.* Minnesota Mathematics and Science Teaching Project. University of Minnesota, 1969.
4. Turtox Service Leaflets. *The Culture and Lab Use of Planaria (#10).* Turtox/Cambosco, Chicago. Illinois. 1974.

Reprinted from *Science and Children*, David C. Kramer, April 1987, pp. 30-32.

Daphnia

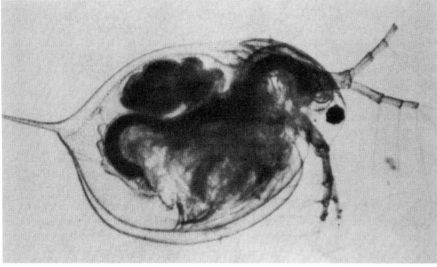

John R. Olive, Alfred J. Hopwood

Daphnia, often known as water fleas, are diminutive aquatic organisms that inhabit freshwater environments throughout most of the temperate areas of North America. Their size, unusual features, and interesting habits make them intriguing to observe—and their role in aquatic food chains makes them valuable tools for teaching certain ecological principles. And, because they're found in abundance in many areas and are easily collected and maintained, daphnia are excellent organisms for short-term classroom study.

Daphnia belong to the class Crusta-

In The Classroom Animal, a development of S&C's popular Care and Maintenance series, column writer David C. Kramer focuses on the natural history of small animals suitable for short-term classroom study and on how to care for these animals. Readers wishing to communicate with Professor Kramer should write him at the Department of Biology, St. Cloud University, St. Cloud, Minnesota 56301.

---

cea, a subgroup of the phylum Arthropoda (the largest group of animals on Earth) that also includes insects and spiders. Also being relatives of shrimp and lobster, daphnia are characterized by an external skeleton, numerous appendages, and gills.

Although they are complex organisms, daphnia are outwardly simplistic in structure. Superficially, they appear to consist only of an oval body (with no appendages) and a head with two prominent features—a single pair of antennae and a pair of eyes. In fact, daphnia have five or six pairs of legs and a short tail, but these are held close to the body and are covered from each side with a fold of the exoskeleton (the carapace)—which obscures the legs and tail and gives the body a clamlike look.

The internal structures of the body—the circulatory, digestive, and reproductive systems—are visible through the light-tan exterior of the daphnia. Details of these structures, even the beating heart, can be observed with a low-power microscope or magnifier. (See figure at right.)

Locomotion

Daphnia are planktonic organisms, meaning that they are weak swimmers and tend to drift with the water's current. But daphnia are slightly denser than water so that, without any active movement, they settle slowly toward the bottom. When drifting in this manner, the daphnia's antennae, which are held above the head, function like parachutes, keeping the tiny creatures upright. Then, by occasionally flipping the antennae rapidly downward, they propel themselves upward, in a jerky motion that makes them appear to be jumping. (This movement resembles that of terrestrial fleas, hence giving daphnia the name water fleas.) Daphnia can control their vertical position in the water by either increasing or decreasing the rate of the antennae movement.

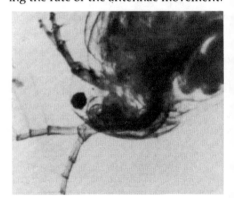

Habitat and Diet

Daphnia can be found in almost any type of freshwater habitat, including streams, lakes, ponds, swamps, and marshes (even those that become dry for periods of time). Interestingly too, the daphnia's habitat to some extent determines its abundance. Larger lakes tend to have low but relatively stable populations throughout the year. In smaller lakes, ponds, marshes, and swamps, daphnia populations have wide annual fluctuations with the highest levels occurring in late spring and again in late summer.

Daphnia feed on a wide variety of microscopic organisms, including algae, protozoa, bacteria, and decaying organic

National Science Teachers Association

material suspended in the water. To gather their food, daphnia generate a current of water with the legs, then filter out the suspended particles with a network of fine bristles on their legs. They then pass the food to the mouth and swallow.

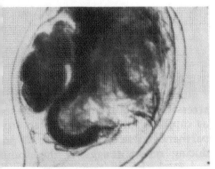

Methods of Reproduction

The most intriguing features of daphnia are their methods of reproduction and how these methods are related to the prevailing environmental conditions. In normal, favorable conditions, a population of daphnia consists almost entirely of females, who reproduce by parthenogenesis (reproduction by development of unfertilized eggs). Each female releases some 2 to 40 eggs into a brood chamber on her back. There the eggs soon hatch and the juveniles are immediately released into the water where they quickly mature and begin to reproduce.

But, should an environment become stressful—if the temperature cools dramatically or the food supply diminishes—each female's offspring will be made up of about half males and half females. These offspring will actually mate and each of these females then produce highly resistant eggs that can tolerate the harsh environmental conditions—even the drying-up of the water. Once the environment returns to favorable conditions, however, the resistant eggs hatch females who return to the parthenogenetic style of reproduction.

Daphnia are prolific. Because each female is capable of producing broods of up to 40 offspring every two or three days for the month or more she lives,

and because each new generation can reproduce within a few days, the population increases rapidly. Naturally, then, in their abundance, daphnia are a major food source for other aquatic organisms, especially fish. As a link between the microscopic food daphnia consume and the larger predators that feed on them, daphnia have an important role in the energy flow of aquatic food chains.

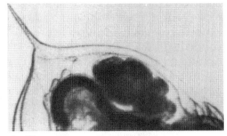

Collecting and Culturing

During periods of high population, daphnia can be scooped up by the hundreds with the single sweep of a fine mesh aquarium dip net. When the population is lower, several slow sweeps of the net through the water in a wide figure-eight motion might be required to accumulate a good many. Once captured, the tiny daphnia will stick to the moist net. To release them, simply turn the net inside out and lower it into a container of pond water into which the daphnia will swim away.

As you collect daphnia, your catch will often include a number of other organisms, such as mosquito larvae and organic debris. To obtain a pure culture of daphnia, selectively recapture the daphnia from the first container by using a large-bore medicine dropper or kitchen baster. Then transfer them to a second container of pond water.

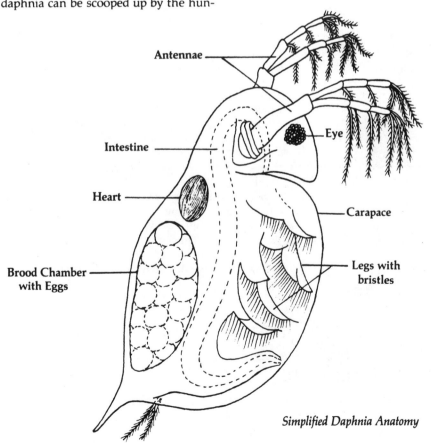

Simplified Daphnia Anatomy

If at all possible, keep the daphnia in the water from which they were taken. To do so, collect several containers of water at the same place you did the capturing. Pour the water through the collecting net into a second container to free it of any other organisms and debris.

In the Classroom

Daphnia can live well in a large (4-L) glass or transparent plastic jar or a standard aquarium. But don't crowd them; one or two dozen daphnia per 4-L container is about right. And, as they reproduce, be sure to divide the culture.

When dividing the culture, if a supply of pond water is not available, use aged tap water, but do mix the tap water with water from the original culture to reduce the environmental shock that can occur when organisms are transferred to a very different environment. Daphnia can tolerate normal classroom temperatures, but they seem to do best below 21° C (70° F), so keep them in a cool part of the room, out of direct sunlight.

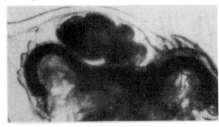

What to Feed Daphnia

Daphnia do well on bacteria, green water (algae), or dry yeast. To produce green water for your creatures, place a

4-L jar of pond water under a bright artificial light or in a sunlit window. Single-celled algae, which occur naturally in the water, will reproduce profusely and soon turn the water green. Alternatively, start an algae culture by placing a teaspoon of garden soil in a 4-L jar of aged tap water that you then keep in the light for a few days. As the green water is poured off (see instructions on feeding that follow), simply refill the container with water and add a drop or two of liquid fertilizer to provide a nutrient source and the algae will continually regenerate.

Culture bacteria by placing one half of a mashed boiled egg yolk in .5 L of water that is then left to stand at room temperature. After two or three days, the water will become cloudy, indicating a high bacteria population level.

If dry yeast is fed, it can be activated in water (much as it is in baking), then fed to the daphnia as described above for feeding them the bacteria.

To feed the daphnia, first siphon or pour out about .5 L of water from the culture, taking care not to lose the organisms. Then add an equal amount of water from the algae or bacteria culture. (This should turn the water slightly cloudy or green.) After two or three days, when the daphnia have cleared the water by consuming the food, simply repeat the process. This method of feeding the daphnia has the dual advantage of providing an adequate food supply and continually replacing some of the water, thus preventing buildup of waste products.

As well, you might add a little mashed egg yolk directly to the daphnia culture where it will decay and produce bacteria for food. This method requires that you replace about one-fourth of the water twice each week to remove accumulated waste material.

Kept as described, daphnia can sometimes be cultured continuously. However, some of the daphnia will eventually die (from age, if for no other reason). Dead daphnia will accumulate in the bottom of the container and should be

left there. The dead specimens will decay and produce more bacteria that the living daphnia will feed on.

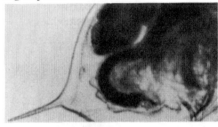

Some Surprises

Remember the oddity of daphnia in stressful environmental conditions too, as captive daphnia often switch from parthenogenetic reproduction to producing resistant eggs. If this should occur, the population will gradually decline as the older daphnia die. But all will not be lost. The accumulated debris will probably contain numerous resistant eggs, giving your students an opportunity for further investigation. Just pour off as much water as possible (without losing the debris) and allow the debris to dry for several weeks. Then add fresh pond water or aged tap water. Though this is pure experimentation, and no results are guaranteed, it is possible that a new generation of parthenogenetic females will develop from the eggs.

Finally, raising daphnia is an excellent cooperative project for groups of four or five students. Each group can have their own daphnia culture and raise their own bacteria and algae for food. By synchronizing the production of bacteria and green water between groups, a continual supply of food will be available for the entire class and if a single culture of daphnia is lost, a subdivision from a stronger one can easily replace it.

Resources

Buikema, Arthur L., Jr., and Sherberger, Sara R. (1977). Daphnia. *Carolina Tips*, XL(10), 37–40.

Pennak, Robert W. (1978). *Freshwater invertebrates of the United States*. New York: John Wiley.

Reprinted from *Science and Children*, David C. Kramer, March 1987, pp. 34-36.

ryptozoa

Earlier articles in this series have dealt with the natural history and classroom care of individual animals. In several cases, it as suggested that an animal be kept in terrarium simulating the natural environment. But of course animals do not ve alone in nature. Rather, they live nd interact in an environment that icludes other animals—herbivores, mnivores, and carnivores; predators, rey, and scavengers—as well as plants nd other components of the natural orld. This article takes a different pproach. It suggests creating a forest oor terrarium as an environment for variety of organisms, a place where nimal interaction can be studied, and a ocal point for learning about a variety f elementary science concepts.

A terrarium is a miniature, enclosed rrestrial environment that simulates natural one. The word *terrarium* is erived from the Latin *terra*, meaning and," just as *aquarium* is derived from *qua* for "water." Since terrariums are nclosed, it is possible to maintain a table environment inside a terrarium hat is quite different from that just utside. Capitalizing on this fact, terariums were first used to house exotic lants requiring a specific environment nd only later came into use as a means f providing an environment appropriate or captive animals. A terrarium can be nade to simulate any type of natural nvironment—desert, bog, marsh, or orest floor—simply by adjusting the

amount of moisture and the type of soil and organisms; and the only limit on the kind of organism that can be housed is the size of the enclosure. But of course a woodland terrarium, based on local flora and fauna, is the easiest to set up and keep in a classroom and therefore the most practical.

Before establishing a terrarium, it is worthwhile to examine, if possible, a similar environment in nature. When we look down on a forest floor, it typically appears to consist of a layer of leaves covering soil. Closer examination will reveal a much more complex situation. The uppermost layer of leaves often includes a few twigs and seeds from the most recent year's growth. Unless there has just been a rainfall, this layer is usually dry; but it protects the underlying layers, which are sometimes several centimeters thick, from rapid evaporation and helps maintain the unique, below-ground environment. By carefully removing the leaves, a few at a time, and examining successive layers, one can observe the conditions immediately below the surface of the ground and the animals that abound there.

Leaves are generally resistant to decay and might require several years to decompose completely. In the first year, there is little decomposition of the dry, uppermost layer. In successive seasons, the

---

In The Classroom Animal, a development of S&C's popular Care and Maintenance series, column writer David C. Kramer focuses on the natural history of small animals suitable for short-term classroom study and on how to care for these animals. Readers wishing to communicate with Professor Kramer should write him at the Department of Biology, St. Cloud University, St. Cloud, Minnesota 56301.

---

soft parts of leaves decompose leaving the skeleton of the veins. Smaller veins decompose next, and finally the larger veins and stems disappear. Twigs and seeds are more resistant, but eventually even these disappear as decomposition continues.

The process of decomposition is accomplished by a variety of organisms including bacteria, fungi, and molds. Herbivores and scavengers—for instance, certain insects, isopods, millipedes, and earthworms—assist in the decomposition by feeding on the partially decayed material. In turn, predators such as ground beetles and centipedes prey on the herbivores and the scavengers. This biological activity, together with the physical mixing of the materials by burrowing insects and worms (especially next to the soil), results in a continuum from top to bottom of increasingly decayed organic material that gradually becomes incorporated into the soil. These processes create horizontal zones or microhabitats, each with its unique assemblage of organisms. Snails, slugs, and isopods occur in the upper layers, centipedes and millipedes are more common in the middle zone, and earthworms in the lower layers.

The accumulation of organic material on the forest floor is sometimes referred to as *duff*. Because this layer has so many tunnels, burrows, and hiding places, it is also sometimes called the *cryptosphere* from the Greek word *kryptein* meaning to hide, and the animals that live here are collectively known as *cryptozoa*—hidden animals. There are many kinds of cryptozoa. Those mentioned above are the most obvious, but others include many kinds of tiny insects, round worms, and spiders. Some kinds of cryptozoa occur in large numbers. Darwin suggested that a single acre of soil could contain as many as 50,000 earthworms, and others have estimated that 1,400 million insects might inhabit that same acre.

One of the reasons for the high population levels in the cryptosphere is the stability of the environment. Although there are significant changes during the annual spring thaw and autumn freeze, the insulation provided by leaf litter means there is little day-to-day fluctuation in temperature. Moisture is of first importance to cryptozoa—nothing is more devastating to them than desiccation—and moisture levels, too, are relatively constant in the cryptosphere.

Light intensity is less crucial to animals than to plants, but most cryptozoa prefer dim light or darkness and, of course, it is always dark under the leaf litter.

A Forest Floor Terrarium

Practically any type of container is suitable for a woodland terrarium, but the larger the container the easier it is to establish and maintain a stable environment. The depth of the container is also critical—it determines the depth of the leaf layer and, therefore, the variety and number of organisms the terrarium will be able to accommodate. A standard aquarium (about 40 L) is a good size, but other containers are also satis-

factory: smaller plastic aquariums, plastic shoe or sweater boxes, or large (about 4 L) glass jars. Whatever the container, it will need a cover to prevent dehydration and to keep the organisms from escaping.

Organisms and substrate can be collected in any woodland area, many vacant lots, and even from under shrubs where leaves have been allowed to

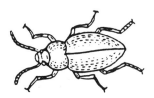

accumulate. Search for larger organisms under the leaf litter or under rocks, pieces of bark, or decaying logs.

To avoid injuring the animals (or being injured yourself), scoop them up with plastic spoons and put them into containers in which they can be carried to the classroom. Avoid over-collecting. One larger organism (like a snail, slug, beetle, centipede, or millipede) or up to five medium-size organisms (like isopods) per 100 sq cm of surface area will be about right. Smaller organisms are difficult to collect singly, but, because of their abundance, they will be included when the soil and leaf litter are collected.

Collect the substrate using a trowel, large spoon, or any convenient tool. Try to get at least three layers (the dry upper layer, the moist decaying matter, and the soil), and keep each layer in a separate container—a plastic bag, for example. (It doesn't matter if materials from a particular layer get mixed together.) Pieces of bark, twigs, moss, small plants, or other naturally occurring material of appropriate size can also be collected to decorate the terrarium and make it look natural.

To set up the terrarium, arrange the substrate in the container, adjusting the thickness of the layers to the size of the container. Since soil holds a lot of moisture and will be useful in maintaining humidity in the terrarium at the re-

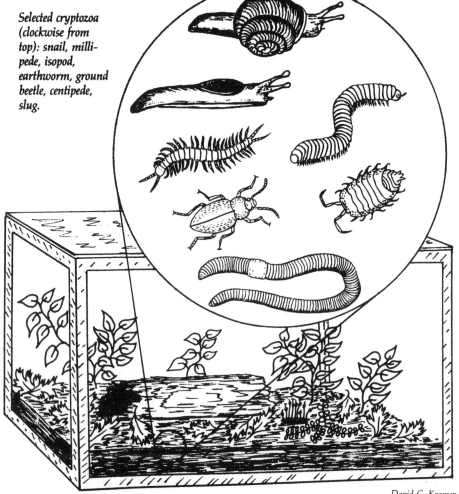

Selected cryptozoa (clockwise from top): snail, millipede, isopod, earthworm, ground beetle, centipede, slug.

David C. Kramer

uired level, make sure that the soil layer is at least 2–3 cm deep. Now add the other materials (bark, twigs, rocks), arranging them to simulate the appearance of a forest floor, and then the plants and animals. Finally, moisten the soil lightly with water. Get students to make a list of the organisms placed in the terrarium.

After the terrarium is established, it will require little care. In the humid environment of the terrarium, the organic material will slowly decay and

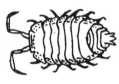

provide food for the herbivorous scavengers. However, some animals—for instance, isopods and millipedes—will consume an occasional slice of potato or carrot hidden under the leaf litter. The tiny soil organisms will reproduce rapidly and provide forage for predators such as centipedes and ground beetles, but if the terrarium has been set up with the proportion of animals to area noted above, the populations should remain roughly in balance.

Since the terrarium will be covered, there will be little loss of moisture, but it might be necessary occasionally to add some water to maintain the slightly humid atmosphere. If moss or other

green plants are included in the terrarium, the water sufficient to sustain them will also be about right for the animal life. Though cryptozoa do not need light, green plants do, so make sure they get enough indirect light. Do not, however, put a terrarium in direct

sunlight: the temperature in the enclosed space could become too high.

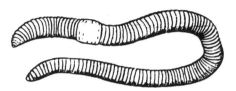

Otherwise, normal classroom temperatures will be satisfactory for the organisms.

Some Teaching Suggestions

1. If possible, have small groups of three or four students each prepare a terrarium. Encourage students to observe their terrariums carefully, daily at first and at least weekly thereafter.

2. Eventually, have students carefully dismantle the terrariums and compare the kinds and numbers of organisms they find to those with which they started. Some kinds may be missing, possibly eaten by predators. Some new kinds might also be found, perhaps having hatched from eggs that were in the soil. And sometimes large populations of certain organisms will be found if reproduction has occurred. Discussing the changes that have taken place will help reinforce students' understanding of a variety of biological and ecological concepts.

Finally, when the project is over, having students release the organisms outdoors in an appropriate environment will demonstrate—and encourage—a positive attitude toward the organisms.

Resources

Broekel, Ray. (1982). *Aquariums and terrariums.* Chicago: Childrens Press.

Lavine, Sigmund A. (1977). *Wonders of terrariums.* New York: Dodd, Mead.

Parker, Alice. (1977). *Terrariums.* New York: Watts.

Silvan, James. (1975). *Pets in a jar: Collecting and caring for small wild animals.* New York: Watts.

Villin, Jean. (1968). *The animal kingdom.* New York: Sterling.

Reprinted from *Science and Children*, David C. Kramer, March 1986, pp. 33-34.

Earthworms

Earthworms, variously called angleworms, garden worms, redworms, fishing worms, and (when they are especially large) night crawlers, are found around the world wherever soil conditions are appropriate. There are many kinds of earthworms. Some are only 3 or 4 cm long when full grown, but one, a native of Australia, can reach a length of 3 m. Some kinds are specific to a certain region or soil type, while others have a wide distribution. Except for size, there are few clues that distinguish one of the common types from another, but it is not necessary to classify an earthworm in order to study it.

Anatomy

Having no internal or external skeleton, earthworms are soft and fleshy. The long, cyclindrical body consists of a series of rings known as *segments*. The segments are progressively smaller toward the head end and, in some species, are somewhat flattened toward the tail end. An enlarged glandular area, the *clitellum*, partially surrounds the body about one-fourth of the way between head and tail. Mucous glands keep the skin moist and give the worm a shiny appearance. Each segment has four pairs of tiny spines called *setae*; two pairs are on the bottom and one on each side. The setae are too small to see but can be felt by holding the worm in one hand and gently sliding it through the fingers of the other.

All of the important internal organs, including the brain, hearts (five pairs of them), reproductive structures, and stomach, are located in the region forward of the clitellum. The mouth is surrounded by the first ring, or segment, which forms the lips. The intestinal tract, a straight tube, extends the length of the body, terminating at the last segment.

Many people mistakenly believe that

This earthworm displays the segmented body and characteristic movement of its kind.

an earthworm broken into two parts will become two earthworms. It is true that, if the body is broken behind the clitellum and not too many segments are lost, the forward portion containing the internal organs will sometimes heal and survive. However, the rear portion of the earthworm, lacking these organs, will die.

Habits

As their name implies, earthworms spend most of their time burrowing in the soil. They burrow by two methods. If the soil is loose, they press their setae against the sides of the burrow to hold their position, then use their nose to push the particles of soil aside. They can also literally eat their way through the soil. Since earthworms need moisture, they will gradually burrow deeper if the topsoil becomes dry. And, as the soil cools in the autumn, they burrow down to spend the winter months below the frostline.

At night, especially when the soil is moist, earthworms come to the surface to consume organic debris and sometimes to mate. They usually extend their forward portions out of the burrow; then, if disturbed, they can quickly withdraw to safety with a few rapid contractions of their muscular bodies. Sometimes, following an exceptionally heavy rainfall, earthworms come to the surface of the ground, apparently seeking drier areas. At these times, they can appear in large numbers on streets and sidewalks, where they sometimes become stranded.

Reproduction

Earthworms are hermaphroditic; that is, each one produces both sperm and egg cells. When mating occurs, each receives sperm from the other, then goes its separate way to complete the reproductive process. Later, the clitellum produces a mucous band that slips forward over the earthworm's body. Eggs and sperm are released into this band as it passes over the reproductive pores. As the band slips off the worm, each end closes to form an oval cocoon. The cocoon remains in the soil for two or three weeks before the eggs hatch. About three months later, the young worms will be fully grown.

Benefits to the Soil

Good soil and earthworms go together and even perpetuate each other. Earthworms prefer rich soil containing abundant organic material, which they consume for food. The organic material also helps keep the soil moist—a necessity for earthworms' survival. In turn, earthworms greatly improve the soil, first by releasing nutrients from the organic materials they consume, and, second, by mixing the soil and keeping it loose. As earthworm burrows aerate the soil, they increase its capacity to hold water. On the other hand, clay or sandy soils that contain little or no organic material and other types of soil that tend to dry out quickly often have low earthworm populations, and, in their absence, seldom improve by natural means.

Charles Darwin estimated that a single acre of soil may contain 50,000 earthworms.

Earthworms are prolific, and in good soil conditions they can reach high population levels. Charles Darwin estimated that a single acre might contain as many as 50,000. In general, the better the soil the higher the earthworm population. However, many other animals, including birds, garter snakes, salamanders, toads, and a number of mammals prey on earthworms. (And even though earthworms do not occur naturally in water, they sometimes become the prey of fish when offered on a hook.)

Caring for Earthworms

An adequate supply of earthworms for classroom study can be found in almost any neighborhood by digging in a garden or overturning rocks, logs, or boards (a good job for some willing students). Worms can also be found by searching the lawn with a flashlight the night after a soaking rain. Or they can be purchased from a bait shop or ordered from a biological supply house.

Once obtained, earthworms can be kept in a variety of ways, depending on the teacher's plans. If they are to be held only a few days, a dozen or two can be kept in a paper milk carton, cottage cheese container, or similar receptacle half filled with moist soil or sphagnum moss. A lid with small holes punched in it will prevent escapes and provide adequate ventilation. It will also help maintain the earthworms at the cool temperatures they favor. (If they are placed in a refrigerator, they will live for several weeks.)

If the goal is a breeding colony, a larger and slightly more sophisticated system will be called for. In this case, the container should be as large as is practical, because the larger the volume of soil, the easier it is to maintain a constant environment. A metal or plastic tub is ideal, but a standard aquarium or a bucket will do. Good garden soil is a satisfactory medium and if mixed with leaf litter, compost, peat, sawdust, or cow manure it will be even better. Place 15–20 cm of the medium in the container and water as needed to keep it moist but not wet. Then two or three dozen earthworms can be placed on the surface and allowed to burrow into the soil. Adding a thin layer of leaf litter or shredded newspaper (avoid paper from colored sections) will help reduce moisture loss.

If the soil is of good quality, the earthworms will not have to be fed for some time, but small amounts of food added to the surface will gradually disappear. As the population increases, some food should be made available at all times. Since earthworms will consume almost any kind of organic debris, they can be fed shredded bits of grass, dried leaves, lettuce, and apple or potato peelings. Earthworms will also eat small amounts of soaked cornmeal, chicken mash, oatmeal, or coffee grounds. A little grass seed sprinkled on the soil will soon sprout and provide a natural-looking environment, as well as a natural moisture meter: if the system is watered enough to keep the grass healthy, the moisture level in the soil will be about right for the worms.

Earthworms will do satisfactorily at room temperature, but their optimum is much lower—about 10–15° C (50–60° F)—so keep the colony in the coolest part of the classroom.

Under good conditions, several hundred earthworms can be produced in a container the size of a small tub in a few months. They can be used in a number of classroom activities. As natural food for many higher animals, a colony of earthworms is also useful as food for other classroom pets.

Resources

Carter, Ian S. (1975). *Invertebrates: Searching for structure.* Toronto: Holt, Rinehart and Winston.

Silvan, James. (1975). *Pets in a jar: Collecting and caring for small wild animals.* New York: Viking Press.

Villin, Jean. (1968). *The animal kingdom.* New York: Sterling.

Snails

Snails belong to an ancient group of animals called molluscs which, along with the octopus, squid, and the familiar clam, originated in the primordial sea. Some marine deposits contain snail fossils that are more than 500 million years old, and these fossils are among the oldest with recognizable living descendents. Thus, the lowly snail, sometimes overlooked and best known for its "snail's pace," is distinguished (if it could speak for its ancestors) by having seen the origin of fish, the invasion of the land by amphibians, the rise and fall of the great dinosaurs, the conquest of the air by birds, and the dominance of the Earth by mammals. During this time, snails have become one of the most diverse and widespread of all animals and are surpassed only by the insects in their variety of living forms. One of their secrets of success is adaptability; snails can now be found in practically any permanent freshwater or marine environment, and some have also adapted to life on land.

Snail Anatomy

Snails are easily recognized by their spiral shells. Since most snails withdraw into their shells for protection, the shell must grow as the snail grows. Thus, it is a spiral of increasing size. There are two styles of shells. In one, the spiral is in a flat plane like a coil of rope on the floor. However, since this kind of shell is difficult to transport, most snails have developed a more compact and easily carried shell with a cone-shaped spiral. Some snail species have solved their transportation problem by developing a smaller shell, and, in some cases, the shell is so small that it no longer serves a protective function. A few forms have carried shell reduction to the extreme and have no shell at all. These snail-like molluscs without shells are called slugs. Some people think slugs are snails that have lost their shells, but this is true only in the evolutionary sense. For those

Reprinted from *Science and Children*, David C. Kramer, September 1985, pp. 41-43.

This aquatic snail displays the conical shell that characterizes most snails and a single pair of tentacles.

Paul Meyers

snails that have a shell, it is a permanent, living, growing part of the body that cannot be abandoned.

The part of the snail that protrudes from the shell is called the foot, but it actually includes most of the animal's body, and its functions comprise sensory perception and eating as well as locomotion. The flat-bottomed muscular part of the foot secretes a thin film of mucous, which provides lubrication, protects the soft body from abrasion, and helps the snail cling to rocks, twigs, and other surfaces as it glides along. The head, which is actually part of the foot, is equipped with tentacles—two pairs in most terrestrial snails and one pair in aquatic ones. One pair, always on top of the head, performs many of the snail's sensory functions. The second and smaller pair, located in the front of the head, would be used for probing the area ahead of the snail. The eyes are either at the base or on the tips of the tentacles on top of the head. The mouth is on the bottom of the head and

contains a rasp-like tongue that is used to scrape off bits of food as the snail moves over surfaces like plants and rocks. Some of the snail's internal organs such as the heart, kidneys, and intestines remain inside the shell even when the snail's foot is fully extended.

Many aquatic snails obtain oxygen through gills, but this system is not satisfactory for terrestrial forms. Land snails take air into a breathing chamber through an opening on the right side of the body just below the shell. This opening can often be seen on land snails when the foot is fully extended. Some freshwater snails also have an air-breathing chamber, so they must occasionally come to the surface of the water for air or circulate oxygen-rich water through the chamber.

Reproduction and Feeding

There are several methods of reproduction in snails. Most terrestrial forms are hermaphroditic. That is, each snail has both male and female reproductive organs. When two of these snails mate, each receives sperm from the other. Then, each later produces eggs, which it lays among leaf litter, under a rock or

The bottom of the snail's foot is visible here as are its eyes (at the base of the tentacles). Between the tentacles is the snail's mouth, or proboscis, which contains its rasp-like tongue.

or in some other hidden place where eggs will stay moist until they hatch tiny snails. Some aquatic snails are hermaphroditic, but others are did into males and females. In at least species, all of the young start life as es, with some becoming females as y mature. Most aquatic snails deposit r eggs in a gelatin-like mass, which ches itself to submerged vegetation ome other available object.

Vhatever their method of reproduc, snails are prolific and would reach n population levels if their numbers e not held in check by predators. iatic snails are consumed in great nbers by many species of birds, inding herons and ducks. Turtles, fish, certain mammals also consume large nbers of aquatic snails. Land snails vulnerable to predation by birds and ious small mammals, including rots and shrews.

Most terrestrial and aquatic snails are narily herbivores, feeding on a wide ety of vegetation including fungi and e. However, they are opportunistic l will feed on the bodies of dead ects, earthworms, and, in the case of iatic snails, dead fish.

uatic Snails

Aquatic snails are among the easiest mals to collect and keep in captivity. nost any permanent body of water ermanently flowing stream will have nail population. However, pond snails better to keep than stream snails ause their habitat is easier to dupli

cate in the classroom. Snails can easily be collected by searching through submerged vegetation or on the surface of any submerged object. The snails should be picked up gently as their shells are sometimes fragile and easily broken. Also, snails usually cling firmly to the substrate by their foot and will be injured if they are quickly pulled off. As soon as the snails are collected, they should be put in a container of the water from which they were taken. One should also collect enough water to fill the container in which the snails will be kept. This will prevent the environmental shock that might occur if the snails were suddenly placed in water of different quality. If possible, also collect a few pieces of the pond vegetation to decorate the snails new home and provide them with a source of food.

Teachers for whom a snail collecting expedition is not practical will find snails and the plants needed for their habitat at local pet stores.

Aquatic snails can be kept in nearly any container that will hold water, but a transparent one will give students a good view of the snails, and one with a wide opening will allow for easy access. A 20-L or 40-L aquarium is fine but so is a large wide-mouthed jar or plastic container as long as the snails are not overcrowded—three to six per liter of

water (depending on their size) is enough. The container can be set up like a typical freshwater aquarium. Put 3–4 cm of sand in the bottom to anchor the plants and provide a natural substrate. After adding pond water to within 8–10 cm of the top, gently push the plants into the sand, and add the snails. A loosely fitting cover will permit air exchange while minimizing evaporation and preventing the snails from wandering out of the enclosure. To keep the plants healthy, place the aquarium where it will receive at least some light each day.

The aquatic snails and their habitat are easy to maintain. The snails will eat small amounts of fish food if provided, but it is not necessary to feed them. Since they are herbivores, they will obtain sufficient food by grazing on the plants. The only regular maintenance needed is to replace the water as it gradually evaporates with fresh pond water or with tap water that has been aged for three or four days. Otherwise, once established, native snails can be kept indefinitely throughout a school year with little effort.

Land Snails

Although adapted to a terrestrial existence, land snails are susceptible to dessication, so their activity is limited by the available moisture. They tend to be active at night when the humidity is higher and are most likely to be seen on the surface following a rainfall. Otherwise, they are usually found among leaf litter, under rocks or logs, or in other damp locations. During dry periods, land snails protect themselves by aestivating, or withdrawing into the shell and sealing the opening with mucous until damp conditions recur.

A wide-mouthed 1-L or 4-L jar, a transparent plastic shoe box, or an aquarium—arranged to simulate the natural environment—is an ideal en-

David C. Kramer

The mucous that these land snails secrete helps them cling to rocks and logs.

closure for land snails (and slugs). The aquarium will need a cover to maintain the necessary humidity level and to keep the snails from escaping, but it will also need some ventilation. A few centimeters of moist soil will help maintain the humidity and provide a medium in which the snails can occasionally burrow and perhaps lay eggs. If the humidity becomes too low, the snails will begin to aestivate. However, they can be encouraged to become active again by adding a little water to the soil.

Captive snails will consume a variety of foods, but a diet of lettuce, carrots, apple, or celery will meet their needs and is easy to provide. An entire carrot can be placed in the terrarium, and, although snails can consume a large amount of food, this will probably last for a week or more. However, land snails can go several days without eating and, if they have adequate moisture, they require no special attention over weekends and holidays.

Resources

Carter, Ian S. (1975). *Invertebrates: Searching for structure*. Toronto: Holt, Rinehart and Winston.

Simon, Seymour. (1975). *Pets in a jar: Collecting and caring for small wild animals*. New York: Viking.

Vallin, Jean. (1968). *The animal kingdom*. New York: Sterling.

Suggested Observations and Activities

- Observe and describe how a snail moves.
- Determine how fast a snail can move.
- Observe and describe snail eggs.
- Find the parts of a snail—the shell, foot, head, tentacles, eyes, mouth, breathing pore.
- If several snails are available, organize a snail race.

Snails and Preschoolers

RUNNING out of exciting science investigations for your preschoolers? Why not try snails? They're easy to obtain, easy to work with, and easy to maintain, according to **Raymond J. O'Toole,** College of Education, University of Arizona, Tucson; and **Ora O'Toole,** University of Arizona. Gather four preschoolers around your science table, add four snails, four magnifying glasses, some lettuce, several 6-in. square pieces of glass, and stand back and observe firsthand learning at its best. Dr. O'Toole guarantees you'll have four very involved, excited youngsters.

Try some of these questions for direction: How does a snail move? What and how does it eat? Does it have a head, feet, or arms? Can it bite? Where does it live? What will it do when touched? How fast does it move? Can it come all the way out of its shell? How does it see? What's that sticky stuff it leaves behind? These questions are actually some of the many questions asked by the children in their exploratory investigations.

Where do you get snails, how do you keep them, and what do you feed them? Snails can be found fairly easily in gardens and lawns shortly after dark or in the early hours of the morning. If you can't find any in your lawn or garden, ask your friends and neighbors if you can look in their gardens. Or try wooded places, beneath rocks or leaves, or along river banks.

A good-sized glass bowl or small aquarium will make a fine home. Add some twigs for climbing on, rocks for crawling under, soil, a fine wire screen over the top, and a bowl of water, and you have a fine home for snails. For food, try lettuce, spinach, fish food, or aquarium plants every couple of days.

For added learning experiences, try some of the following: Assign various members to be snail-caretakers for a week; as an art project, make snails out of paper and/or clay; take dictation from the children about their experiences and observations with the snails; let various children dramatize the movements of the snails.

Two very good sources of information concerning snails can be found in: *Snails* by Dorothy Hogner (Crowell, 1958), and *A Book of Snails* by Sally Kellin (Young Scott Books, 1968).

Reprinted from *Science and Children* October 1973. p. 6.

Terrestrial Isopods

Many children are familiar with the terrestrial isopods (crustaceans) commonly called pill bugs, sow bugs, or wood lice. Terrestrial isopods are easily collected and kept in classrooms. Isopods are useful for many scientific investigations by elementary students.

Their gray to blackish colored bodies are oval, flat, and 5 to 15 mm long. The sow bug or wood louse (species of *Porcellio*) has two tail-like appendages and a flattened body that let the animal right itself easily from an upside-down position. These isopods do not roll into a spherical shape but only flex their bodies slightly. Some isopods (species of *Armadillidium*) have a high domed body. They can roll into a spherical shape which gave them the name pill bug. (See Figure 1) Isopods may roll into "pills" when they are disturbed or perhaps to lessen water loss by evaporation when humidity is low.

You can trap isopods with a cut raw potato.

Land isopods live in dark, moist places beneath undisturbed objects lying on the ground—rotting logs, boards, bricks, or rocks. Sometimes isopods can be found alongside buildings if there is enough moisture and a food source for them.

Finding and collecting isopods by using popsicle sticks to separate grass and other vegetation from the base of a building makes a good class field trip. Put the isopods in a container along with soil and vegetation to take back to the classroom. Do not close the container tightly. Another collection method is to put a raw hollowed-out white potato near a likely source of isopods, cover it with dead grass or leaves, and leave the trap for a day or two. (See Figure 2) If isopods are present, they will be attracted inside the raw potato trap.

Care

To keep classroom isopods, put 2 to 3 cm of rich woods soil or commercial potting soil in a plastic sweater or shoe box, terrarium, empty aquarium, or other container. (See Figure 3) Provide a darkened area by partially covering the soil with some material that allows for the free circulation of air. For ex-

Reprinted from *Science and Children*, Carol D. and Carolyn H. Hampton, March 1979, pp. 60-61.

Figure 1. Shapes of isopods.

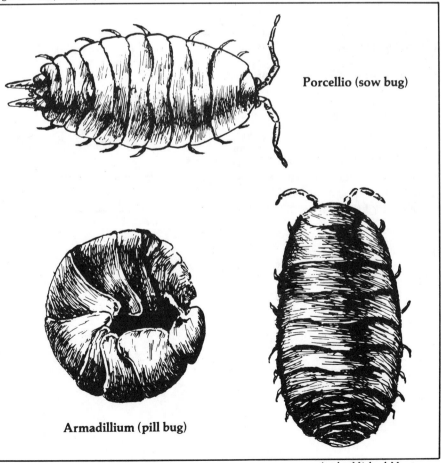

Porcellio (sow bug)

Armadillium (pill bug)

Art by Michael Montonara

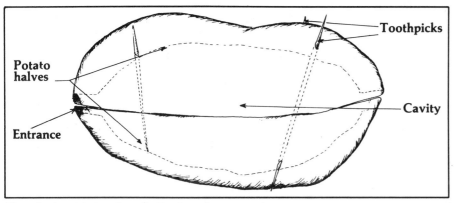

Figure 2. A potato trap for collecting isopods.

ample, support a piece of corrugated cardboard on a few pebbles or put a section of paper egg carton over half the soil. You can add a few pieces of decaying bark or wood. Sprinkle the container with water making sure the habitat is damp but not saturated. Put some isopods in the container. Once a month add a slice of raw potato. Alternate the potato diet with small pieces of raw carrot, lettuce, unsweetened breakfast cereal, and ripe fruit. Remove food that begins to mold. Cover the habitat with a lid that allows ventilation. If you see water standing in the soil, remove the lid and allow the excess water to evaporate. Once the habitat is set up there is little maintenance needed other than feeding and watering.

Isopods are prolific and will produce many young in three to four weeks. It is hard to sex isopods but a fertilized female is identified by the eggs carried in a brood pouch called a marsupium. The marsupium can be seen on the underside of the female between the legs of the third through the sixth body segments. Hatched young remain in the marsupium a few days before they are

visible in the habitat. They are pale immature replicas of the adults.

Isopods make good laboratory animals for demonstrating structure-function relationships and responses to changing environmental conditions. Isopods can be raised as food for toads, frogs, and lizards. Students can develop experiments to test isopods' responses to light intensity, color, food

preferences, temperature, and humidity changes. The resources include sources of activities as well as collection and maintenance information.

Resources

Barnes, Robert D. *Invertebrate Zoology.* Third Edition. W.B. Saunders Co., Philadelphia, Pennsylvania. 1974. Pp. 430-434.

Behringer, Marjorie P. *Techniques and Materials in Biology.* McGraw-Hill, Inc., New York City. 1973. Pp. 76-79.

Hunt, John D. "Race Them, Race Them." *Science and Children* 14:26-27; January 1977.

"Isopods." *Outdoor Biology Instructional Strategies.* Trial Edition, Set III. University of California, Berkeley, California. 1977.

Kramer, David C. "Isopods: The Un-Bugs." *Science and Children* 11:14-15; January/February 1974.

Morholt, Evelyn, et al. *A Sourcebook for the Biological Sciences.* Second Edition. Harcourt, Brace and World, New York City. 1966. Pp. 599-600.

Figure 3. Habitat for isopods.

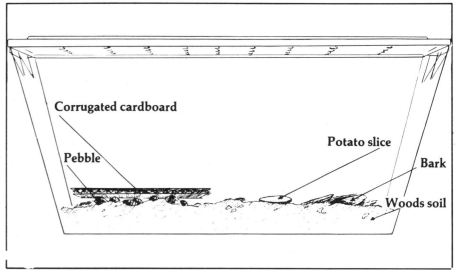

Crayfish

Over three hundred kinds of crayfish occur in North America. Each has specific habitat requirements and is thus limited in its range. However, since there are so many species, crayfish can be found in almost every imaginable aquatic or semiaquatic environment throughout southern Canada and the U.S.

Crayfish—sometimes called *crawfish, crawdads, mudbugs,* or *crabs*—are freshwater organisms related to marine shrimp and crabs, and, of course, to lobsters, which they closely resemble. All are members of a group (or class) called Crustacea, characterized by a hard exoskeleton, numerous legs, and the presence of gills. These, in turn, are members of a larger group, the phylum Arthropoda, which includes insects and spiders and is the largest group of animals on Earth.

Characteristics

Although they vary in length from 5 to 12 cm or more, most species of crayfish are so similar in appearance that it is difficult to distinguish between them. Usually reddish to brown or grey-brown in color, their bodies are divided into two parts: a rigid *cephalothorax* and a flexible, segmented *abdomen.* Two pairs of *antennae,* a pair of eyes on stalks, and the mouth are located near the front of the cephalothorax, and five pairs of legs are attached to the midline along the bottom. The first pair of legs terminates in large pincers—the most obvious and characteristic feature of the crayfish. These are used in eating, jousting with other crayfish over territorial matters, defense, and burrowing. The four smaller pairs of legs are used primarily for walking but are also tipped with small claws that help them cling to the substrate as a crayfish climbs. Most of the internal organs, including the brain, heart, lungs, and stomach, are in the

Some crayfish burrow to the water table where they create underwater chambers.

cephalothorax. The abdomen looks and functions like a tail, but it too has small leglike appendages called *swimmerets* and a broad, flat, fanlike projection called the *telson.*

Walking is the chief method of locomotion for crayfish. Considering their cumbersome appearance, crayfish are surprisingly agile and can move forward, backward, or sideways in water or even on land, where they sometimes wander. They can also swim short distances, but they are awkward swimmers and adopt this way of getting around mostly as a means of escape. They swim by making a series of downward and forward movements of the abdomen, which propels them backward in a jerky fashion.

Most crayfish live among rocks and litter on the bottom of streams, rivers, swamps, marshes, ponds, lakes, and muddy backwaters, and in these habitats they often reach high population levels. Other, solitary, species burrow in fields and meadows, sometimes as deep as a meter, to create underground chambers just below the water table. The entrances to these burrows are often "chimneys" of mud balls, the remains of the process of excavation. (See figure.) A few species called *cave crayfish,* unique because they are colorless and eyeless, inhabit underground streams sometimes hundreds of meters below the surface of the Earth.

Diet

Omnivorous and opportunistic in their feeding habits, crayfish play an important role in many aquatic ecosystems. First, as scavengers, they consume a great deal of dead plant and animal material that would otherwise decay. They also consume living plant material and, as predators, they eat insects and their larvae, worms, snails, fish, frog eggs, and tadpoles. When they eat, crayfish hold food against their mouths using one or both pincers, rasp away tiny bits, and then swallow.

In turn, crayfish are preyed on by many other animals including various

Reprinted from *Science and Children*, David C. Kramer, January 1986, pp. 126-128.

fish, turtles, snakes, mink, raccoons, and a number of birds such as kingfishers.

Reproduction

Reproduction in most species takes place in late summer or fall. A few weeks after mating, females produce an average of 100 to 200 (but up to 400) dark-colored eggs. Once laid, the eggs are attached to the swimmerets and carried about by the female. When carrying the eggs, the female is said to be in the "berry stage."

After hatching, the young, which resemble miniature adults, remain attached to the swimmerets for a week or two. Then, when they are about 1 cm in length, they leave the female and lead an independent existence.

Molting

The young grow rapidly and molt several times before they reach maturity in as little as three to four months. Thereafter, both males and females molt two times each year—once to produce the breeding stage and once to produce the nonbreeding stage—during their 2–4 year lives. Crayfish can regenerate an appendage that has been severely injured or broken off. The process begins with the next molt, and eventually a new leg or pincer appears.

For a few days following each molt, the new exoskeleton, including the greatest defense of the crayfish, the pincers, is soft and flexible. At this time, crayfish are especially vulnerable to predation, so they tend to be reclusive until the exoskeleton hardens. They are sometimes collected during this "soft-craw" stage for use as fish bait.

A Delicacy

Considering their kinship to lobsters, shrimp, and crabs, it is not surprising that crayfish are edible. They are not widely exploited for food except in certain southern and West Coast states and a few other areas, where they are commercially harvested and consumed in great numbers. Peeled tails (abdomens) fried in butter and whole crayfish boiled like lobster are reputed to be excellent.

Housing, Care, and Capture

Crayfish are territorial and will fight if crowded. This characteristic can be the source of interesting behavioral studies, but it needs to be considered in maintaining captive specimens. One or

two crayfish can be kept in a 40–L aquarium. If more are kept, they should either be given separate accommodations or a larger container such as a plastic wading pool. Two to five cm of coarse aquarium gravel will provide an appropriate substrate, and a few rocks, a brick, or a clay flower pot will provide hiding and climbing places. Crayfish have a seemingly uncontrollable urge to alter their surroundings. They will continually move the substrate around, and they often burrow under objects. Then they will defend the special area they have created from other crayfish. Crayfish need a only few cm of water, but there should always be enough to completely cover the specimens.

Captive crayfish can be fed any of their natural foods. But, since these are sometimes unavailable and since crayfish are omnivores and scavengers, a variety of substitutes will also be acceptable. Small pieces of lettuce and other vegetables as well as pieces of fresh meat or fish will be an adequate diet.

Crayfish are messy eaters, and it will be easier to keep their cage clean if they are removed to another container (a bowl or pail) for feeding. Otherwise, any uneaten food should be removed from the enclosure after half an hour or so to prevent fouling of the water.

Though crayfish can be taken out of the water for observation and study, their tolerance for dryness is limited since they breathe through gills. Consequently, they should not be kept out of the water for more than 10 to 15 minutes at a time. Crayfish adapt well

to life in the classroom and can be kept successfully as long as two years.

Handle crayfish carefully. They will not bite, but they can give a painful pinch with their strong claws. To pick them up safely, grasp them on either side of the cephalothorax above the walking legs with the thumb and forefinger. This technique should be demonstrated to children if they are to be allowed to handle crayfish.

Crayfish for classroom use can often be collected from shallow streams. Place a large net downstream from a rock, and gently move the rock. This will alarm any crayfish sheltering under the rock, and as they attempt to escape, they will be swept into the net by the current. Crayfish can also be caught, with practice, by raising a rock and grabbing them.

Or try fishing for them. Tie a string to a piece of meat or fish and dangle it near a crayfish. When the crayfish grasps the bait with its pincers, gently raise it out of the water and lower it into a pail. No hook is needed as the crayfish will hold on and catch itself.

Crayfish are protected in some areas of the U.S., so the propriety of collecting them, if in question, should be determined by contacting a local conservation officer. Alternatively, they can often be purchased from fishbait shops or from biological supply companies.

Resources

Crayfish: Teacher's guide. (1976). Elementary Science Study. New York: McGraw-Hill.
Pennak, R. W. (1978). *Freshwater invertebrates of the United States.* New York: John Wiley.

This crayfish, pincers at the ready, also displays the four pairs of walking legs and segmented abdomen terminated by the fan-shaped telson.

Jeff Korte

Reprinted from *Science and Children*, David C. Kramer,
October 1987, pp. 24-26.

Daddy

David C. Kramer

Longlegs

lthough harvestmen, or daddy longlegs, are widespread and often abundant, there are many misconceptions and misunderstandings about them. Even the source of their name is a mystery. They are reputed to be called harvestmen because they are often seen in profusion at harvest time when they

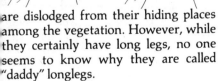

are dislodged from their hiding places among the vegetation. However, while they certainly have long legs, no one seems to know why they are called "daddy" longlegs.

Daddy longlegs are sometimes erroneously called "spiders" but,

although related and somewhat spider-like in appearance, harvestmen are not true spiders and have little in common with them.

Spiders have two distinct body regions (the cephalothorax and abdomen, but

the harvestman body is not divided into distinct regions. Harvestmen also lack poison glands and fangs, they have only two eyes rather than several, and do not produce silk—the hallmarks of spiders. Spiders and harvestmen are in different orders within the same class, *Arachnida*, along with scorpions, mites, and ticks.

There are many kinds of harvestmen.

Some are small and indistinct and others even have short legs.

About 200 kinds of the familiar and frequently encountered forms of the long-legged harvestmen occur throughout the United States and parts of Canada. These types are the focus of this article.

In The Classroom Animal, a development of S&C's popular Care and Maintenance series, column writer David C. Kramer focuses on the natural history of small animals suitable for short-term classroom study and on how to care for these animals. Readers wishing to communicate with Professor Kramer should write him at the Department of Biology, St. Cloud University, St. Cloud, Minnesota 56301.

The harvestman body is oval or egg-shaped and, depending on the species, is typically 4–8 mm in length. Usually light to dark brown, some species are reddish or black. The mouth is located on the underside of the small end of the oval body, and the eyes are on either upper side near the front. Harvestmen have four pairs of legs, all of which are used for walking, but the second pair are longer than the others and are also used as feelers for probing the environment. The males of a given species have proportionally longer legs than their female counterparts, but the females have larger bodies for the production of eggs.

Habitat and Natural History

Harvestmen are adaptable wanderers, never having a permanent home or hiding place. Their nature haunts include fields, meadows, roadsides, and woodland areas where they may be found in or on the ground cover and among the vegetation. They are also commonly found in lawns and around the foundations of buildings and nearby vegetation as well as inside garages and basements. The requisite for their occurrence seems to be the presence of appropriate food and moisture.

Harvestmen are primarily predators and as such feed on a wide range of tiny insects, mites, spiders, and other similar organisms, including each other. While harvestmen require some moisture, their needs are minimal and they seem to be able to obtain enough water from their food and the morning dew and can thus be found in very dry places.

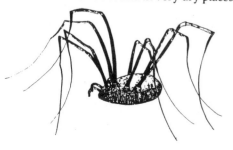

Being slow-moving and nonagressive, harvestmen are vulnerable to other animals such as birds that search for food among the vegetation. However, harvestmen have several passive ways

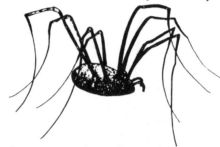

of protecting themselves. Their color, shape, and slow movements provide protection through camouflage—protective coloration, resemblance, and behavior. Also, when disturbed, harvestmen produce a strong and (presumably) offensive odor that discourages predation. Finally, if other systems fail and a predator happens to catch a harvestman by a leg, the fragile appendage is easily broken off to allow the owner to escape—the leg is later regenerated.

Life Cycle

After mating in summer or early fall, female harvestmen lay their eggs in the soil or among the leaf litter of the forest floor. The eggs hatch the following spring into tiny immature harvestmen that, apart from having proportionally shorter legs, closely resemble the adults. Once hatched, the young assume a nomadic life style much like the adults and consume similar foods. They grow quickly and in the process, molt several times and become increasingly more adultlike in appearance. By midsummer, the young are fully grown and ready to mate and produce the eggs that will after winter become the next year's harvestmen population.

In all but the warmer areas of their range, harvestmen succumb to the cold with the onset of winter and thus live less than one year. In the warmer areas adult harvestmen may hibernate for

short periods; they retreat to a hiding place and their metabolic rate slows as the temperature drops.

Housing and Study

Since harvestmen do not have poison glands or fangs and will not bite humans, they can be handled safely. However, since they are so fragile (their legs easily broken), it is in their best interest not to attempt to capture or hold them by hand. They can easily be captured for short-term observation and study for a few hours by quickly, but gently, placing a transparent jar or similar container over them and then sliding a lid or piece of cardboard underneath.

A single harvestman can be confined in a jar or similar container with a hole punched in the lid for ventilation. A leaf, twig, or other piece of vegetation will provide a climbing or resting place

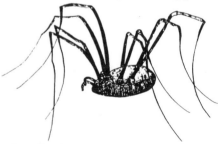

and make observations easier and perhaps more interesting. Add a few drops of water for sufficient humidity to prevent dehydration.

For a longer period of confinement, harvestmen can be kept in a forest floor terrarium. (See "Cryptozoa," *S&C*, *24*(6), 34–36.) Here a 5 or 10 gallon terrarium is arranged to simulate a forest floor and includes a few centimeters of soil, some decaying leaves, and a variety of soil organisms. When the soil and leaf litter is collected from the forest, it will contain a rich variety of organisms, cryptozoa. Assuming that sufficient moisture is provided, these organisms will reproduce and provide a source of food for the harvestmen as well as for any other predators in the terrarium.

National Science Teachers Association

If two or more harvestmen are kept in the same container, one might consume the other, especially if they differ greatly in size. Remember, no captive animal should be placed in direct sunlight; the heat can quickly build up to dangerous levels.

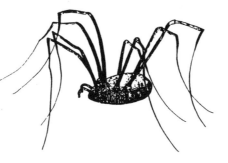

MANY TEACHERS raise organisms in their classrooms. From the biology lab that is half zoo to the primary classroom with goldfish or a gerbil, animals have proven themselves a good means of attaining cognitive as well as affective educational goals.

Spiders can add new interest and learning opportunities to collections of living classroom animals. They are easy to collect on field trips and are easy to keep throughout the year. The learning that students can experience by observing and caring for the life needs of these organisms can be enhanced by several simple activities which demonstrate basic life science principles. Here are some tips for raising and caring for these organisms.

Spiders are easier to catch and care for than most organisms. Collect them from open fields in the fall or spring, or from your home or school any time of the year. If you can collect many, choose different species, perhaps some that come from different environments or that spin different kinds of webs. This way you can compare eating habits, growth rates, reproductive habits, or behavior among species.

For a spider home, all you need is a relatively large glass jar with a lid, some soil and a few twigs. A large pickle bottle or a gallon mayonnaise or catsup jar will suffice. (Large spiders may require larger jars.) Spiders eat each other, hence, only one to a jar. Place about 5 centimeters of slightly damp soil on the bottom and prop some twigs along the sides upon which the spider can build its home. Punch air holes in the lid and put the spider inside.

Spiders can eat houseflies, fruit flies, wood lice, mealworms, or even crickets. As long as you remember to feed them something about their own size or a little smaller every 2 weeks, they will thrive. If your spiders do not respond to this bi-weekly schedule, observe their eating habits and modify your feeding accordingly. Spiders obtain moisture from two sources: their prey and the atmosphere kept humid by the damp soil. Therefore, when the soil begins to dry, moisten it slightly with a light spray of water.

Observing Harvestmen

Children can make many observations of the structure and behavior of harvestmen. Encourage them to compare harvestmen to insects and spiders; observe and describe how harvestmen use their legs for walking and as feelers; determine the three ways harvestmen protect themselves by camouflage; and, if students observe carefully, they might be able to see a harvestman capture and consume its prey.

Resources and References

Farrand, John, Jr. (1986). *The Audubon Society encyclopedia of animal life.* New York: Portland House.

Kramer, David C. (1987, March). Cryptozoa. *S&C*, 24(6), 34–36.

Levi, Herbert W., and Levi, Lorna R. (1968). *A guide to spiders and their kin: A golden nature guide.* New York: Golden Press.

Palmer, Lawrence E., and Fowler, H. Seymour. (1978). *Fieldbook of natural history,* (2nd ed.). New York: McGraw Hill.

Spiders As Classroom Pets

Spiders that spin orb webs must be fed live insects on the web. They will not take prey off the ground. Hunting spiders, such as wolf spiders and jumpers, take live food off the bottom of the jar.

Many different activities can be done with these organisms. Spiders can be used to show a few basic ecological interactions. Put some in a terrarium, possibly with crickets.

The food pyramid, predator-prey relationships, the noncyclic nature of energy flow in the community, and other ecological concepts can be introduced through the interactions observed. If an organism dies, it begins to decompose and ultimately returns to the soil, showing that decomposers play a large part in the community cycles and balance.

Among the simplest of activities are those that involve feeding. Instead of telling the students what to feed their new pets, let them find out. Will the organisms eat vegetables? Peanut butter? Raw meat? Live insects? Or even bologna sandwiches? Questions of this nature when posed by and to your students can be answered by making a chart, trying the different foods, and observing the outcome. In addition, these uncomplicated feeding questions should soon cause other more interesting ones to arise, such as "How do you know if the spider ate your food?" or "How much food did the spider eat?" These are not simple questions to answer; they will certainly lead the students to do some very creative thinking.

Reproduction and maturation also provide an interesting focus for student investigations with topics such as birth conditions, birth rate, and life span observable for study. The necessary physical conditions for birth can be investigated by isolating a spider's egg sac. One can then systematically manipulate the physical variables of interest, such as heat and moisture, to discover optimum birth conditions. Subsequent observation of the organisms can lead students to determine averages for the number of young produced under these conditions, the number reaching maturity, and the life span of the survivors.

Other imaginative activities can be done with spiders if you follow the students' questions and interests.

Bibliography

1. Bason, Lillian. *Spiders.* Books for Young Explorers. The National Geographic Society, Washington, D.C. 1974.

2. Couchman, J. Kenneth, et al. *Small Creatures.* Holt, Rinehart and Winston of Canada, Limited, Toronto, Ontario. 1974.

3. Dupre, Ramona Stewart. *Spiders.* Follett Publishing Company, Chicago, Illinois. 1967.

Reprinted from *Science and Children*, Padilla, Michael J., September 1977, pp. 11–12.

Reprinted from *Science and Children*, David C. Kramer, May 1986, pp. 42-44.

Centipedes and Millipedes

Lifting a rock, board, or log will often reveal a variety of invertebrates including insects and their larvae, spiders, worms, isopods, centipedes, and millipedes. The last two animals are vaguely similar and often confused. Both are multilegged organisms that share a habitat and a tendency to be secretive. Both are occasionally unwelcome guests in people's houses. Even their names reinforce the tendency to confuse them. In fact, centipedes and millipedes are vastly different, and, since they are easy to capture and maintain in the classroom, the misconceptions about them can be readily dispelled. Moreover, students learning about how to distinguish one from the other can also study a variety of ecological principles.

Habitat and Characteristics

Centipedes and millipedes are found from the tropics to the subpolar regions of the world. Both require a humid environment, and in temperate regions they generally live among leaf litter and other decaying vegetation. Both are primarily nocturnal and tend to hide (under rocks, boards, pieces of bark, or other available objects) during the daytime. And both have similar reproductive patterns—first mating, then laying their eggs in the moist soil where they hatch without further attention from the adults.

There are certain physical similarities, as well. Both centipedes and millipedes have an elongated body consisting of numerous leg-bearing segments, a rigid exoskeleton, and (like their arthropod relatives such as insects and spiders) jointed legs. Beyond this, there are many differences between them. Although they share a common habitat, they fill two completely different ecological niches: millipedes are herbivores and feed on the decaying vegetation in which they live; centipedes are predators and feed on isopods, small insects, and other small invertebrates. Another difference is their shape. Millipedes have a rounded body, and their short legs protrude from below; centipedes are flattened, with longer legs that extend to either side of the body.

Then there is the matter of the legs and segments. The Latin prefixes *centi-* and *milli-* mean, respectively, 100 and

Centipedes and millipedes don't necessarily have 100 or 1,000 legs as their names imply—the most legs ever recorded on a millipede was 750.

1,000, and *pede* comes from a Latin word meaning foot. But centipedes and millipedes do not necessarily have 100 or 1,000 feet (or legs) as their names imply—one European centipede does have up to 177 pairs of legs, but the most legs ever recorded on a millipede was 750 (Chinery, 1984). In both animals, however, the number of legs is directly related to the number of body segments. Centipedes always have one pair of legs on each leg-bearing segment; millipedes have two pairs on each. The number of segments on most centipedes or millipedes depends on two factors—age and species. Most have only a few segments when they hatch, and they add segments with each molt as they grow. Some species have as few as 10 segments; others have over 100. Thus, if a centipede happened to have 50 segments, it would have 100 legs and be correctly named. A millipede, on the other hand (or leg), would need to have 250 segments to have 1,000 legs; but, as already noted, none reach this length.

In keeping with their feeding habits and physical characteristics, centipedes and millipedes have different levels of

A millipede on the forest floor, where it finds the decaying vegetation on which it feeds.

Paul E. Meyers

Paul E. Meyers

The centipede's flat body and long legs help it move quickly to catch its prey.

activity and means of defending themselves. Centipedes' flat bodies and long legs allow them to move quickly through the leaf litter as they use a pair of venomous pincer-like claws on the head to catch and immobilize prey. If centipedes are disturbed, their primary means of defense is to run away and attempt to hide under some object. But, if captured or detained, they will attempt to bite.

The herbivorous millipedes, on the other hand, are slow moving and passive. Their first line of defense is to coil into a spiral. If further disturbed, they will release a foul-smelling substance from glands located in each segment. Both centipedes and millipedes are said to be poisonous, but this term is relative. Yes, centipedes can poison their tiny prey, and the caustic substance secreted by millipedes might poison a predator such as toad, but neither organism, except for a few tropical species, is harmful to human beings and, even with these species, the "poison" is more irritating than toxic.*

Occasionally a centipede or millipede will wander into a house or accidentally be brought in with something like a load of firewood. Such an appearance is more of an annoyance than a problem. Neither organism is harmful to the occupants or the contents of the house. Nor, since both normally live in decaying leaf litter, are they likely to set up quarters or propagate indoors. However, if provided with an appropriate environment, both organisms do well in the classroom.

Housing and Care

A terrarium with a forest floor environment is suitable for keeping both centipedes and millipedes. Several millipedes can be kept in the same enclosure because they are herbivores and will not bother each other. And their caustic secretion means that the centipedes won't bother them either. But, unless centipedes have an abundant food supply, they will prey on each other—especially if crowded together—and eventually their population will dwindle to one.

Practically any container in which a suitable environment can be maintained

is a satisfactory enclosure, though a transparent glass jar, plastic shoe box, or an aquarium make it easier to observe the inhabitants. Since maintaining a stable environment is important, a large container (with a lid and a few air holes) is probably the best choice.

Centipedes and millipedes can be purchased from a biological supply company, but they can also be found and collected in their natural environment by gently lifting rocks or logs. Plastic spoons make good tools for hesitant collectors and are less likely to injure the specimens than fingers, another standard tool. When collecting the centipedes and millipedes, it is a good idea to collect a few

*A check with other experts in the field, prompted by reader response to this article in S&C, led to a clarification in the Nov./Dec. '86 issue. When children handle millipedes—or any other classroom creatures—see that they keep their hands away from their eyes, and wash the hands thoroughly afterwards. Since some southwestern centipedes inflict painful bites, and some individuals react drastically to the venom of even the lesser species, centipedes should only be handled in a plastic box or vial that is taped shut.

Paul E. Meyers

of the other small organisms, especially isopods from the same site. (See Kramer, 1974, January/February, and Hunt, 1977, January.) These organisms will make the terrarium more natural and the isopods will, as they reproduce, provide a continuous food supply for the centipedes. Also, collect some soil, leaf litter, and other natural objects such

---

Most forest floor organisms do not drink water but, rather, absorb what they need from the surroundings.

---

as pieces of bark, a small log or stump, or rocks for use in the terrarium.

On returning to the classroom, place about 5 cm of the soil in the container; then arrange the leaf litter and other objects to simulate the natural environment. Finally, add the organisms.

Once established, the terrarium will be easy to care for. The most important consideration is that the soil be kept moist, which is no problem if a cover is provided. Most forest floor organisms do not drink water but, rather, absorb what they need from their surroundings, so drinking water is not needed. And, since the millipedes (and isopods) will feed on the decaying leaf litter and the centipedes on the young isopods and other small organisms, feeding is

A ventral view of a millipede exhibiting its numerous leg-bearing segments. Each of these segments has two legs (rather than one, as with the centipede).

not necessary either (though an occasional slice of raw potato will ensure that the isopods and millipedes always have a ready food supply). Eventually, the populations will balance out and, if properly maintained, the terrarium, with a few centipedes and several millipedes, can be kept for several months.

Because of their secretive natures, centipedes and millipedes will be easier to observe if removed from the terrarium and temporarily placed in pill vials, petri dishes, or other convenient transparent containers. Encourage students to observe them carefully. How many segments does each have? How many legs? How are they alike? How do they differ? Also, centipedes and millipedes can be used as examples of, or to illustrate, various ecological concepts such as predator–prey, carnivore–herbivore, habitat, niche, and adaptation.

Resources

Barns, Robert. (1963). *Invertebrate zoology.* Philadelphia: W. B. Saunders.

Chinery, Michael (ed.). (1984). *Dictionary of animals.* New York: Arco.

Hunt, John D. (1977, January). Race them, race them! *S&C, 14*(4), 26–27.

Kramer, David C. (1974, January/February). Isopods: The un-bug. *S&C, 11*(5), 14–15.

Palmer, Lawrence E., and Fowler, Seymour. (1975). *Fieldbook of natural history* (2nd ed.). New York: McGraw-Hill.

Lepidoptera

Studies of the complete metamorphosis in insects are readily made in the classroom by rearing one of the members of the Lepidoptera. Various species of butterflies or moths may be collected as gravid females ready to oviposit or as recently laid eggs. The eggs in either case may be placed in small containers such as pill boxes, jars or finger bowls, and covered with a top that will allow circulation of air. When the eggs hatch, the larvae may be placed in larger containers and fed daily on fresh food plant material. The food plant offered should be that specific for the species. If this is not known or cannot be determined from the literature, use the species of plant on which the eggs or adults were collected. Plant material that has wilted or dried out should be discarded and replaced with fresh material.

Many common species of Lepidoptera have more than one brood a year and therefore pupation and emergence of the adults may be observed. However, production of fertile eggs under laboratory conditions is usually not feasible. (Wards Natural Science Establishment, Inc.)

Reprinted from *Science and Children*, March 1989, pp. 29-31.

Mealworms
in the
Classroom

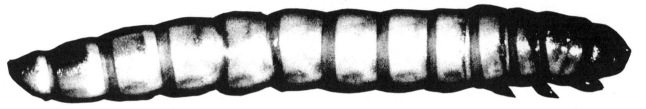

This readily available insect has all the best qualities for a life-cycle study in the classroom.

A re you looking for a classroom animal that is inexpensive, clean, easy to care for, safe, and interesting? Look no further, for the larva of the darkling beetle—commonly known as the mealworm—fits the bill.

There are about three million insect species, and one million of these are beetles. In fact, beetles comprise a full quarter of all species in the animal kingdom. Although the darkling beetle, or *Tenebrio molitar,* is fairly undistinguished among the beetles, it meets most of the requirements we set for a classroom animal and thus is the ideal beetle for classroom study.

Buy 'em by the Box

This magnificent specimen is easily acquired. Since the mealworm is a source of food for carnivorous pets such as chameleons, lizards, turtles, and frogs, pet stores sell them in lots of 25, 50, or 100 for about two to four cents per animal, depending on the quantity purchased. A box of 25 is small enough to carry away in your pocket and provides students with hours of fascinating investigation.

Once you have purchased your new class pets, transfer them to a ventilated, non-cardboard terrarium. You will need to add a little more food since the small supply that accompanied the mealworms is usually insufficient. Any dry cereal will work, either flakes, oatmeal, or bran.

Except for providing water, you have already done almost everything necessary to care for the mealworms. A small slice of apple, carrot, or raw potato added weekly will provide enough moisture for the larvae. *T. molitar* spends most of its life in the larval stage, typically six to eight months depending on the temperature. Keeping the mealworms refrigerated will slow or temporarily halt the cycle. [For a more detailed discussion of housing, care, and feeding, see Kramer, 1985.]

By Glenn McGlathery

The short yet dramatic life cycle of T. molitar *makes it ideal for elementary classroom study. After several months in the larval stage (left), the mealworm pupates and spends the next few weeks as a comatose pupa (below) before emerging as an adult (right) that will quickly mate to begin another cycle.*

Observing

Mealworms are about 2–3 cm long, so students should use a magnifying glass to make observations. Start your class by distributing a mealworm in a Petri dish or a plastic cup to each pair of children. Ask them to make as many observations as they can about the animal. If they are squeamish about handling the mealworms, assure them that they do not bite and are not slimy.

Don't give any more instructions until you feel that students have had sufficient time to observe. A few minutes later, have them share their observations with the rest of the class. Make sure that observations cover color, length, diameter, number of legs, number of segments, and method of locomotion. Once they have shared ideas, give your students time to make observations that they missed on their own. Then distribute a list of questions and activities to guide inquiry and encourage independent experimentation (see figure).

Some students will notice that their "worm" has six legs. This finding may signal something significant to many students—an *insect* has six legs, but this doesn't *look* like an insect. With further observation, they will discover that the mealworm is simply an insect in its larval stage, the second of the four metamorphic stages an insect undergoes in a complete metamorphosis from egg to larva to pupa to adult.

Awaiting the Transformation

It is now time to make some longterm studies of *T. molitar*. Give each team of students a small, clear plastic pill vial, a small amount of the food medium from the classroom terrarium, and two or three mealworms. Have the students make careful observations every two or three days and note how the mealworms change.

Students may initially notice evidence of molting, the process by which the mealworm sheds its skin as it grows.

You will probably find these shed skins in your original purchase of mealworms, but by the time the mealworms are sold in stores, they usually have reached their maximum size, and little molting occurs until the transformation into the pupa stage.

This final larval molt is dramatic. Its body color darkens before molting, and major body changes begin to occur. The mealworm becomes shorter and wider, its head enlarges, and the 13 body segments begin to fuse together. The color then lightens to an almost transparent white before it yields to its comatose pupal form, in which it will remain for one to three weeks.

To a child, the metamorphosis seems like magic, but it's only the magic of science.

On Becoming Adults

The class should continue making observations on a regular basis. Point out the changes in head structure and the fusion of the 13 segments to form the characteristic three segments (head, thorax, and abdomen) of the adult insect. One final molt will occur, when the adult emerges complete with nonfunctional, vestigial wings. Nonetheless, you will be able to observe some preflight behavior, so look for wing extension and flapping. Are your students able to determine from which original segments the wings generated? Did the mealworm give any signs that this would occur? Did observation of the pupal stage yield a clue? Note the gradual change in the color of the adult from light tan to the solid black that is characteristic of *T. molitar*.

If you put the adults back into the original terrarium, they will mate and the females will lay about 500 eggs and die soon thereafter, having had an adult life of about one month. The eggs will hatch after one or two weeks, and the cycle will begin again.

With four to six weeks of observations, your class will usually witness all four metamorphic stages of the darkling beetle, although it may take as long as 12 weeks if the larvae are young when purchased.

To a child, the metamorphosis seems like magic, but it's only the magic of science. The mealworm demonstrates that it is the ultimate transformer, changing from a long, wormlike, many-segmented animal to a comatose white pupa to a black, three-segmented, streamlined beetle. There are few such opportunities for genuine observation and investigation that are so rewarding and yet so easy.

Resource

Kramer, D. (1985). The classroom animal: Mealworms. *Science and Children, 22*(4), 25–26.

Glenn McGlathery is a professor of education at the University of Colorado at Denver. Photographs by Bruce Thomas.

Questions About Mealworms

Body
How many segments does a mealworm have?
Do all mealworms have that many?
Do mealworms grow?
What is their size range?
How much do they weigh?
What is their range of color?
Describe their body texture.
Does their body temperature depend on room temperature?
How many antennae do mealworms have?
Where are the antennae?
How long are the antennae?
Is there a standard ratio between body length and diameter?

Legs and Movement
How many legs does a mealworm have?
On which segments are the legs?
How long are they?
How do mealworms move?
On which surfaces do they move best?
What is the steepest incline they can climb?
How fast do they move? Have a mealworm race.
Can they walk on a string?

Behavior
Do mealworms congregate?
What temperature do they prefer?
Do they prefer light or darkness?
Do they have a favorite color?

How do mealworms eat?
What do they eat?
How much do they eat?
How do they grow?
How do they eliminate waste?
Do they move toward food?
Do they move toward water?
Do mealworms have intelligence?
Can they solve a simple maze?

Activities
Draw a mealworm
Write a story about a mealworm.
Write a haiku poem about a mealworm.
Write a play about a mealworm.
Be a mealworm. How do you feel about people asking all these questions about you?

Crickets

Of the several kinds of crickets that live in North America, the black field cricket is the most familiar. It lives in grassy fields, vacant lots, gardens, roadsides, and lawns. Another, the house cricket (sometimes called the gray cricket), is brown or gray in color and lives in similar places, but it is somewhat less common. Except for the difference in color and the house cricket's slightly longer wings, these two species are similar in appearance. A third kind, the cave or camel cricket (which is actually not a true cricket), resembles the other two but has a distinct hump on its back. The cave cricket favors cool, moist, dark places and is generally found under rocks and logs.

Crickets live in cracks in the ground, in small chambers they dig in the soil, or under objects such as rocks or boards. Their food is mostly green plant material or some plant derivative such as fruits or seeds. They need some moisture but usually obtain enough from their food. Because they prefer warm temperatures, they are most active during late summer and early fall; and, being primarily nocturnal, they are more often heard than seen.

The chirp of a cricket is an insect equivalent of a bird call: its primary purpose is to attract a mate or mark the cricket's territory. Only the males call, and the sound is made by rubbing the wings together (not the hind legs as is sometimes thought). The front wings have rough, rasplike surfaces that, when rubbed rapidly together, produce the chirping sound. Crickets perceive this sound, and others, with "ears" located on the inside of their front legs. (The ears consist of membranes that function like an eardrum to sense vibrations.)

Since the cricket is an ectothermic, or cold-blooded, organism, its metabolism, and therefore its rate of calling, are affected by the ambient temperature. As a result, one can make a rough approximation of the temperature in degrees Fahrenheit by counting the number of chirps every 15 seconds and adding 40. To calculate the temperature in degrees Celsius, one would divide the number of chirps in 1 minute by 7 and add 4.

Reprinted from *Science and Children*, David C. Kramer, November/December 1985, pp. 28–30.

Paul E. Myers

Note this cricket's antennae (at right) and its cerci and ovipositor (at left). The spiky protuberances on its legs are used for defense and sometimes locomotion.

(Crickets are trying to change from English to metric, but they are presently operating under both systems.)

Anatomy and Reproduction

Crickets show the typical insect attributes. They have three body parts: head, thorax, and abdomen. The head is equipped with eyes and antennae for sensing the environment and a mouth. Two pairs of wings and three pairs of legs are attached to the thorax. The abdomen contains the reproductive organs and most of the digestive system.

A row of small holes on the thorax and abdomen called *spiracles* are the openings to the respiratory system. All crickets have two projections about one-half the length of the body extending rearward from the abdomen. These are called *cerci* (singular *cercus*) and are used to detect vibrations. Females have a third, longer projection called the *ovipositor* between the cerci.

After mating the female pushes her ovipositor into the soil and releases a single egg; this process may be repeated over 2,000 times in her brief lifetime. Crickets exhibit gradual (incomplete) metamorphosis. Each egg hatches into a tiny nymph, which only superficially resembles an adult. The nymph molts several times, each time becoming more like an adult in appearance. With the final molt, the cricket becomes fully developed and sexually mature. The entire life cycle requires from two to four months depending on the temperature. Warmer temperatures speed the process.

Most adult crickets do not survive the winter, so the size of the summertime population is largely a function of the eggs that overwintered in the soil. The population reaches its peak in the autumn as more and more crickets hatch. And, because the length of the average life cycle is temperature-dependent, the higher the average summer temperature, the larger the cricket population.

In nature, crickets are preyed on by many animals, including birds, toads, and insect-eating snakes. Human beings have also learned to use crickets for a variety of purposes. In China and Japan, crickets are sometimes kept as house pets for the pleasure of hearing their song. They are excellent fish bait and pet food for lizards and larger aquarium fish. Crickets are also widely used in schools to study ecology, animal behavior, physiology, and entomology. House crickets are usually used for these purposes as they are easier to raise in captivity than field crickets.

Keeping Crickets

Crickets are easy to keep in the classroom—if one is prepared to tolerate a little chirping. Their needs are simple, and they can be kept in two ways depending on the outcome desired. If the goal is to keep them for a short time, a covered jar is satisfactory. A breeding colony requires more space and a little

additional care. In either case, field crickets or house crickets can be used, but house crickets are superior for rearing and breeding. Live crickets can either be collected from their environment or purchased from a local fish bait dealer or a pet shop. Or they can be ordered from a biological supply company.

To keep a cricket for a few days, place 2–4 cm of sand or soil in a jar with a ventilated cover. Add a dry leaf or a crumpled paper towel and the enclosure is complete. The soil will provide a medium in which the cricket can dig and the leaf, a surface on which it can climb and a place to hide. Crickets will eat a wide variety of foods, but a slice of apple, carrot, potato, or celery or a piece of lettuce is a good short-term food as it will also provide the cricket with the moisture it needs. The food should be replaced every day or two so it will not decay or mold.

A Breeding Colony

A breeding colony of crickets can be kept in much the same way, except that a larger container is needed and two seemingly inconsistent requirements have to be met. First, the environment must be kept dry to prevent disease. And, second, the crickets must have moist soil or sand in which to lay their eggs. Meeting these two requirements is the real secret of raising crickets.

A standard aquarium is an excellent container for housing a breeding colony. The crickets cannot climb the smooth glass walls, and the sides are tall enough to prevent them from jumping out—but a screen cover is a good idea. Again, add 2–4 cm of sand or soil. Crumpled newspaper will provide a hiding place and a surface for climbing, but empty egg cartons with a few holes punched in them are better: if the cartons are put into the aquarium open side down, the crickets will climb inside and out. Crickets will consume almost anything including, if they run out of food, each other. Dry dog food is a well-balanced diet and is easy to provide, but crickets can also be well nourished on oatmeal, corn flakes, bran, or any other grain cereal. Any dry food, however, should be supplemented occasionally with leafy or succulent vegetables. Water should be available continuously, but crickets will fall into an open container and drown. The best way to provide water is to invert a small jar or vial of water in a shallow dish with a

few thicknesses of paper towel between the jar and the dish. The crickets will be able to get their water from the moist towels.

A small colony of 20–40 females can lay several hundred eggs each day, and these can be collected in a shallow dish of moist sand or soil (a plastic margarine tub is perfect). The dish should be slightly recessed in the sand so the crickets can climb in. Adult crickets will eat some of the eggs (and will also cannibalize newly hatched young), so the dish should be removed after a few days and replaced with a fresh one if more eggs are to be collected. Put the dish with the eggs in another escape-proof container (a small aquarium or plastic shoe box) for incubation and hatching. It is essential that the eggs not dehydrate, but this can be prevented by lightly sprinkling the soil if needed. The eggs will hatch in three or four weeks depending on the temperature. The tiny nymphs will come to the surface of the soil and out of the dish. At this time they can be fed, watered, and cared for just as the adults are. After about three weeks they can be kept with the adults with less risk of their being eaten.

Here are a few additional tips. Keep the breeding colony warm. Crickets will survive at room temperature, but they will be considerably more active and will reproduce better at higher temperatures —30°–35° C is about right. A light bulb in the cage is a good heat source. Keep the cage clean. Remove accumulated droppings, any dead crickets, and uneaten food. And remember, keep the colony dry.

Paul E. Myers

Reprinted from *Science and Children*, David C. Kramer, April 1986, pp. 30-32.

Fruit Flies

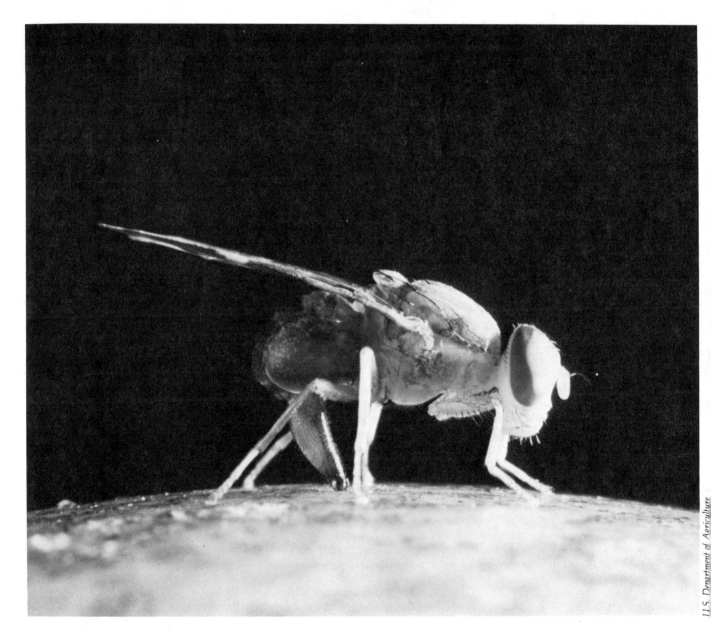

Fruit flies, sometimes called vinegar flies, drosophila (after their scientific name *Drosophila melanogaster*), or, erroneously, gnats, are the diminutive insects that appear around fruit bowls or stored fruit. Since fruit flies are also attracted to fruit peelings and cores, they are often found around garbage cans and compost piles. And, of course, they are common in fruit producing areas, where they feed and reproduce on overripe or decaying fruit. There are several kinds of fruit flies, all similar in appearance.

A female Caribbean fruit fly, her ovipositor tipped downward, prepares to continue her productive life by laying eggs in a piece of fruit.

Characteristics

Although small—about 3 mm in length—and difficult to observe with-

56   National Science Teachers Association

out a microscope or magnifier, fruit flies are similar to larger flies and show most of the typical insect attributes. They have three body parts: the head, thorax, and abdomen. The head, dominated by two large, red eyes, has a pair of barely perceptible antennae and the mouth parts. Three pairs of legs and a single pair of wings are attached to the thorax, which contains the heart and part of the digestive system (though most of

reduced to about 10 days.

Fruit flies reproduce in astonishingly high numbers. A female fly, which can mate during the first day of her life and begin laying eggs the next day, can produce as many as 500 eggs in her first 10 days; and she continues laying, though at a slower rate, for as long as one month. A standard text on insects (Borror et al., 1981) spells out the implications of fruit flies' fecundity:

> If all the offspring of a pair of drosophila survived and reproduced for one year, with the original and each succeeding female laying 100 eggs before

under control.

Fruit flies are best known for their role in the study of heredity. Since they have a short life cycle, a high reproductive rate, and several easily identifiable characteristics, such as eye color and wing shape, it is possible to trace a single genetic trait through several generations in a relatively short time. As a result, fruit flies have probably contributed more to our knowledge of heredity than any other organism.

A female fruit fly, which can mate during the first day of her life and begin laying eggs the next day, can produce 500 eggs in her first 10 days.

the thorax is filled with muscles that move the wings and legs). The abdomen contains the remainder of the digestive system and the reproductive structures.

Fruit flies are generally yellowish in color with several black bands across the abdomen. Females and males can be readily distinguished by examining their abdomens with a magnifier. In male fruit flies, the tip of the abdomen is rounded and has a broad black band; in females, it is slightly more elongated and has a narrow black band. Also, the male's abdomen has five segments but the female's has seven. These differences become more obvious when the female's abdomen becomes distended with eggs. Another distinguishing feature of the male (and one that requires still higher magnification to observe) is the "sex comb," several bristles found on each front leg.

Life Cycle and Reproduction

Fruit flies undergo complete metamorphosis, so there are four distinct stages in their life cycle: egg, larva, pupa, and adult. The length of the life cycle and of each stage is directly related to the temperature, with warmer temperatures tending to increase the rate of development. For example, at 20° C, the entire life cycle takes about 15 days, but at 25° C, the time required is

she dies and each egg hatching and growing to maturity. . . the twenty-fifth generation [would consist] of 1.192×10^{41} flies. If this many flies were packed tightly together, 1,000 to a cubic inch, they would form a ball of flies, 96,372,988 miles in diameter— or a ball extending approximately from the Earth to the Sun.

Wow! Fortunately, fruit flies don't reproduce at this level. Under natural conditions, the fruit fly population is usually low following the rigors of winter, but as the spring and summer progress, and as more fruit is available on which the flies can reproduce, their number continues to increase until a seasonal high is reached in the autumn.

Fruit flies can be annoying when they appear around a bowl of fruit, but most kinds of fruit flies are harmless. They do not have biting mouth parts and they have not been implicated in the transmission of any disease. And, since they feed and reproduce primarily on decaying fruit, they do no harm to crops.

However, Mediterranean and Caribbean fruit flies and three or four other kinds of fruit fly, none of which belongs to the drosophila family, could pose a serious danger to crops. These species lay their eggs in unripened fruits, and the developing larvae (maggots) can destroy an entire crop before it can be harvested; but agriculture officials have thus far been successful in keeping them

Raising Fruit Flies

Fruit flies are easy to raise in the classroom. So easy, in fact, that last autumn several generations of them lived and reproduced in our classroom jack-o-lantern, and we became aware of their presence only when someone bumped the jack-o-lantern and a cloud of flies emerged from the eyes, nose, and mouth. Most teachers, though, will probably prefer to raise the flies under more controlled conditions and in enclosures. Drosophila kits, available from science supply companies, include everything that is needed—containers, media, and flies—in classroom quantities.

Fruit fly photographs on this page by John Coulter.

However, the flies can also be cultured using readily available materials and, as indicated by our experience with the jack-o-lantern, flies are not hard to find—they will find you.

To trap wild fruit flies, put a few pieces of ripe, or even decaying, fruit such as apple, cantaloupe, or banana in open jars and place the jars outdoors. A spot near garbage cans where the flies tend to congregate, while not necessary, will help assure a good catch. It should take two or three days to establish the colonies, and then one need only place a cover (a piece of cloth secured with a rubber band) on the jars and bring them indoors. Flies hovering around the jars are a good sign that eggs have been laid inside, but even if no flies are visible, it is likely that some will have visited the traps by this time. Keep these jars at room temperature, and the flies will reproduce and become the "stock" cultures from which students can establish smaller colonies.

Students' colonies can be housed in test tubes, plastic pill vials, baby food jars, or any similar containers. The size and shape of the containers is not important, but they do need covers that will both prevent the flies from escaping and allow air to enter. Because it provides a consistent environment and will get neither moldy nor smelly, commercial medium is probably the most convenient food source for fruit flies, but they can also be raised on instant mashed potatoes if a mold inhibitor such as calcium propionate is used (Coulter, 1968). Or they can be raised on various kinds of fruit, preferably very ripe or even partially decayed and cut or mashed so the flies have access to the juices. (A little yeast sprinkled on the fruit will speed up fermentation and promote fly reproduction.)

Prepare to transfer flies from the stock culture to students' colonies by sharply tapping the jar on the table to shake the flies to the bottom. Then, quickly replace the cloth cover with a 3 x 5 card in which one or more holes have been made with a paper punch. If students invert their containers over the holes, the flies will crawl in and when 10 or 12 flies have entered, students can quickly cover their containers and their cultures are complete.

Students should observe their colonies each day. The eggs, which the flies will lay on the mashed potatoes or the cut surfaces of the fruit, will be difficult to see because they are so small. If students look carefully, they should be able to find some larvae after three or four days, though even these will be difficult to see at first as they are small and spend most of their time burrowing through and consuming the fruit. Eventually, after four or five days, the larvae will crawl out of the fruit and change into pupae. These, then, will soon change into adult flies, and the life cycle will have been completed within 10 to 15 days, depending on the temperature.

Resources

Borror, Donald J., et al. (1981). *An introduction to the study of insects* (5th ed.). Philadelphia: Saunders College.

Coulter, John C. (1968). Instant potatoes as a fruit fly culture medium. *School Science and Mathematics, 68*(4), 57.

Demerec, M., and Kaufman, B. P. (1964). *Drosophila guide.* Washington: Carnegie Institute.

Flagg, Raymond O., and Noah, Linda J. (1970). Drosophila mutants. *Carolina Tips, 33*(13), 49-51.

Palmer, Lawrence E., and Fowler, Seymour. (1975). *Fieldbook of natural history* (2nd ed.). New York: McGraw-Hill.

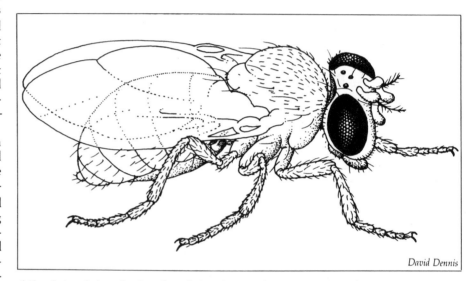

David Dennis

(Above) A male fruit fly. Its red, multifaceted eyes and tiny antennae are characteristic of Drosophila melanogaster. *(Below) A Mediterranean fruit fly laying eggs in an orange. Unlike most fruit flies, medflies pose a danger to crops.*

Paul L. Meyers

Reprinted from *Science and Children*, David C. Kramer, September 1986, pp 46–47

A praying mantis, with spiny forelegs in position to seize its prey, peers over a leaf.

the segments, and, in the case of most flying insects, a pair of wings is attached to the second and third segments. Though mantids share most of these features, their forward thoracic segment is greatly elongated and is attached to the middle segment by a flexible rather than a rigid joint. This difference produces the features that most people associate with mantids—a greatly elongated and flexible body, with the first pair of legs attached far forward and adapted for grasping and holding prey rather than walking. The mantid's second and third thoracic segments still provide a firm base for the attachment of wings (yes, mantids can fly—although awkwardly) and two pairs of legs, which are the insect's primary means of walking and climbing. Another characteristic—and one that is shared with only a few other insects—is that mantids are able to move their head from side to side, a feature that is useful as they search for prey.

Because the mantid's forelegs are adapted for seizing prey rather than locomotion, they are typically held in a folded position. This has been anthropomorphically described as *praying* (hence this mantid's name). However, since the forelegs thus folded are all set to reach out and grasp prey—insects and their larvae, spiders, and even other mantids—the name *preying mantis* might be more appropriate.

Finally, mantids may be unique among insects in having a single ear located on the underside of their body between their legs.

Eating and Mating

Mantids seek their food either by slowly and quietly moving through the vegetation or by lying in wait for passing prey. Their keen eyesight, flexible body, and ability to reach out quickly with their strong forelegs allow them to capture even flying insects. Mantids hold their catch securely with their forelegs and quickly consume it, often eating only the soft parts and leaving behind wings, legs, and pieces of exoskeleton.

After mating, which occurs in late

Praying Mantises

Praying mantises, sometimes simply called mantids, rank among the most fascinating of North American insects. The combination of their large size, striking body form, and intriguing habits makes them especially interesting to keep and study in the classroom.

Many kinds of mantids are native to the tropics, but several also occur in North America. North American mantids tend to be either light brown or green—colors that allow them to blend in with foliage and thus hide from their prey. Like other insects, mantids have a body that is divided into a head (which sports a pair of large compound eyes); a thorax (the point of attachment for three pairs of legs and two pairs of wings); and an abdomen (the location of most internal organs, including the circulatory, digestive, and reproductive systems).

Mantids differ from most other insects, however, in the way their forward thoracic segment is attached to the second segment and in the way it is used. The thorax of an insect typically consists of three rigidly joined segments. One pair of legs is attached to each of summer or early fall, the female ordinarily lays from 200 to 300 eggs in a frothy mass that is attached to some vegetation—often a weed or twig. The frothy material quickly hardens into a firm, ovoid egg case, which resembles a piece of gray or tan foam insulation and is about 2 cm across. When the young mantids emerge from the egg cases in the spring, they have no wings, and they are less than 1 cm long. From the outset, they are highly predacious and are voracious eaters. As a result, they grow quickly. Since the hard exoskeleton limits their growth, they must periodically shed it (molt) as they increase in size. By late summer, they will be up to 10 cm long, and, at the time of the final molt, they develop wings, become sexually mature, and mate to produce the eggs that will become next year's mantid population.

After they mate, the females, continuing their predatory ways, often consume their mates. One might look on this as the male's final contribution to the continuation of the species. But the females themselves don't live long after the laying of their eggs. Adult mantids do not overwinter, so, with the coming of cold weather, the female mantids—and any males that have lived beyond the mating process—die.

Mantids have a reputation as consumers of other insects, so gardeners are often glad to see a mantid in their garden. Sometimes they even buy mantid cases from garden catalogs in the hope that hungry young mantids will help reduce the insect population. However, since the young gradually disperse from the place where they are hatched and many are lost to predators such as birds (and other mantids), the value of this practice is difficult to assess.

But whether or not buying a mantid case will keep a garden pest free, it is a good way of obtaining mantids for classroom study. This is especially true now that the mantid population seems to have declined in parts of their range—due perhaps to pesticides and perhaps to the disappearance of the shrubs and other vegetation that are their natural home. And even where mantids are still numerous, their protective coloring and secretive nature make finding one a matter of luck.

Housing and Care

An adult mantid can be kept in a large jar, small aquarium (if a screen cover is available), or even a wire cage. Since young mantids can crawl through a screen, a glass container with a cloth cover is needed if one plans to hatch an egg case. And, since mantids will prey on each other no matter how small (or large) they are, each mantid will need its own container.

The humid environment that is important both for developing eggs and newly hatched mantids can be provided by placing some slightly moist soil in the bottom of the container. (Prevent egg cases from touching the moist soil by propping the twigs to which they are attached against the wall of the container—and, if a case is loose, attach it to a twig or stem with a drop of white glue.) Something to climb and

A flexible joint between the mantids' forward and middle thoracic segments allows these insects great maneuverability in hunting and capturing prey.

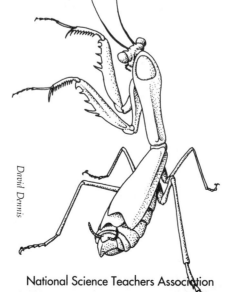

rest on is also an important part of a mantid's environment, and one or more twigs will do the job.

As soon as the young emerge, they should either be released in a suitable environment or separated. Otherwise, all the offspring of a pair of mantids (as many as 1,000 young) may be reduced to a single specimen in a short time.

To feed newly hatched mantids, provide them with very tiny insects such as fruit flies. The insects can be captured or raised and placed in the mantid cage. Or bring a young mantid to its food by placing it in a fruit fly culture.

An adult mantid can be fed any of its natural foods, which includes almost any kind of insect or larva as long as it is alive. Since mantids in nature capture their prey in vegetation, they might have difficulty with ground dwelling insects. If so, offer food with forceps or even with your fingers.

Finally, all captive mantids should be provided with water each day. Put a few drops inside the cage—perhaps on the end of a twig—or try encouraging the mantid to take water from your finger.

A Note on Handling

Mantids are secretive and difficult to find, but if you find one, it will be fairly easy to catch. If possible, maneuver the mantid into a jar or other container. A mantid can also be picked up, but it is fragile and can be easily injured unless handled gently. Also, although its bite is not poisonous, a mantid can and will bite, especially when it is frightened, and the spines on its legs can inflict a painful wound.

Resources

Conklin, Gladys. (1978). *Praying mantis: The garden dinosaur.* New York: Holiday House.

Johnson, Sylvia A. (1984). *Mantises.* Minneapolis: Lerner.

Orlans, F. Barbara. (1977). *Animal care from protozoa to small mammals.* Menlo Park, Calif.: Addison-Wesley.

Palmer, Lawrence E., and Fowler, H. Seymour. (1978). *Fieldbook of natural history.* (2nd ed.). New York: McGraw-Hill.

Simon, Seymour. (1975). *Pets in a jar: Collecting and caring for small wild animals.* New York: Viking.

Reprinted from *Science and Children*, David C. Kramer,
November/December 1986, pp. 31-33.

David C. Kramer

Walking Sticks

Walking sticks occur both in the temperate and the tropical regions of the world. Some of the tropical species are notable for bizarre shapes and for colors resembling leaves or brilliant tropical flowers. Those that occur in North America are typical "stick insects" like the ones pictured in the photographs on these pages. Their range is the U.S. and southern Canada east of the Rocky Mountains, and they are occasionally found in the Southwest.

Although walking sticks are best known—and have been named—for their resemblance to twigs, these gentle insects have several other qualities that make them interesting to study and easy to keep in the classroom. Walking sticks' twiglike appearance is their major means of defense, and almost everything about them, from their shape and color to their habitat, diet, and behavior, is consistent with this mimicry. Few other organisms demonstrate so well the complementarity of structure, function, and behavior.

Walking sticks are related to crickets and grasshoppers but, as seems evident from their appearance, they are even closer to mantids. Like other insects, walking sticks have three body parts—head, thorax, and abdomen—three pairs of legs, and a single pair of antennae. Most North American walking sticks are wingless. Thus, walking is their only means of locomotion. When first hatched, they are typically green in color, but as they mature, some gradually change to various shades of brown. Males are slightly more slender than females and have a small, pincerlike projection on the tip of their abdomen. And, as adults, walking sticks reach an impressive length of 8–10 cm.

Camouflage

Walking sticks' camouflage, their principal line of defense, is of three types. *Protective resemblance* is the most obvious: a walking stick is shaped like a stick, with a head that is small and indistinct and a thorax and abdomen that are very long and slender. The illusion of twigginess is further enhanced by long, slender legs and antennae, which look like the smaller branches on a twig. *Protective coloration*, another form of camouflage, complements the shape—walking sticks are either green or some shade of brown, the color of most twigs. The walking stick's *protective behavior* represents a third form of camouflage. The insects move slowly and deliberately and tend to "freeze" at the slightest disturbance, thus protecting themselves from predators and enhancing their resemblance to twigs.

Some walking sticks, for whom camouflage may not be enough, also protect themselves by giving off a smelly chemical.

Habitat and Reproduction

Deciduous trees and shrubs provide the habitat, food and drink, and protection for walking sticks. The insects feed on leaves—seemingly preferring oak leaves although they also eat those of maple, basswood, willow, and many other trees. Leaves also satisfy their needs for moisture. And, of course, the trees provide protection as well as a place to mate.

Walking sticks mate and lay eggs in the autumn. Each female produces

Paul E. Meyers

National Science Teachers Association

This walking stick's resemblance to the twig on which it perches is its first line of defense against its predators.

hundreds of tiny brown-and-white eggs, which are dropped from the trees and scattered over the forest floor. On the cool ground, walking stick eggs have a long incubation time, and most do not hatch until the spring following their second winter. Young "sticks," looking like tiny adults, then climb into the foliage, where they grow and molt several times before becoming adults. Walking sticks do not survive the cold weather in places where freezing temperatures occur.

For the most part, walking sticks, which do not bite or make noise, are innocuous as well as inconspicuous. Occasionally, when favorable conditions lead to an unusually large population, walking sticks are more in evidence and can even damage trees by defoliating them.

A good method for collecting a specimen is to carefully search through the foliage of trees or gently sweep the leaves with an insect net. Sticks—and especially their legs—are fragile, and they must be handled carefully to avoid harming them. The safest way is to encourage them to walk into a jar or one hand by guiding them with the other.

Housing and Care

Keep a single walking stick in a wide-mouthed jar with a ventilated cover and one or more sticks in a covered aquarium or screen-covered cage. A twig will give the walking stick a climbing and resting place, and one or two fresh leaves, preferably from the tree where the stick was collected, will provide both food and water. In a larger cage, a leafy twig in a vase of water will keep the residents in food and water for several days.

To confine a walking stick outdoors, place a fine net bag (like some of the laundry bags used for washing fragile items) or a leg cut from a pair of panty

hose over a twig with several leaves on it, and tie both openings around the twig. With this technique, the only limitation on the time a walking stick can be kept is the number of leaves in the enclosure. It is important, though, to place the net cage where it will be shaded and where it will not be bothered by animals or curious human beings (who might be particularly likely to investigate if they spy the panty hose).

Hatching Eggs

A captive female walking stick will sometimes produce eggs. If so, they can be collected for hatching and put in a ventilated jar with 4–6 cm of slightly moist soil. Eggs kept at room temperature do not require the long incubation period of eggs that remain outdoors and will hatch in 8 to 12 weeks—probably in the middle of winter when there will be no tree leaves to feed the young—so it's a good idea to delay the hatching. This can be done by keeping the eggs in a refrigerator until about February. Then the young walking sticks will hatch at about the same time as the new leaves appear, and they can be kept in the way already outlined for the adults.

Resources

Evans, Howard E. (1984). *Insect biology.* Reading, Mass.: Addison-Wesley.

Hegner, Robert. (1937). *Parade of the animal kingdom.* New York: Macmillan.

Palmer, Lawrence E., and Fowler, H. Seymour. (1978). *Fieldbook of natural history.* (2nd ed.). New York: McGraw-Hill.

In The Classroom Animal, a development of S&C's popular Care and Maintenance series, column writer David C. Kramer focuses on the natural history of small animals suitable for short-term classroom study and on how to care for these animals. Readers wishing to communicate with Professor Kramer should write him at the Department of Biology, St. Cloud University, St. Cloud, Minnesota 56301.

The Reigning Monarchs

Butterflies in the classroom, from start to finish.

Reprinted from *Science and Children*, April 1988, pp. 12-14.

By Madalyn Brown and Jeffrey R. Lehman

In two months, classes will be dismissed and bulletin boards dismantled. September and beginning the next school year may be the farthest things from your mind. So, think instead of butterflies. Monarch butterflies. They arrive in September with the milkweed pods and the next generation of eager first graders.

Hatching monarchs just may spark a lifelong interest in science for young children. If you remember to collect milkweed and caterpillars at the same time as you're redoing your classroom, the first week of school can also be the first week your new students begin using science processing skills.

Monarch butterfly eggs are difficult to find and identify—they're only 7 mm. long. But the monarch caterpillars are easily recognizable, because you'll find them eating milkweed leaves. For the most part, you will have to collect the animals yourself. Scientific suppliers don't sell them.

At the beginning of the school year, make a science corner with an aquarium in your classroom. Set a small container of water inside of it, along with several milkweed plants and monarch caterpillars. To keep the caterpillars from escaping, cover the top of the aquarium with a piece of screen. Make the screen into a lid by framing it with strips of wood.

During the first few weeks of school, add fresh milkweed leaves and a few more caterpillars to the aquarium. Keep up the supply of food until all of the caterpillars have changed into chrysalides. The screen will prove an excellent place for the caterpillars to attach themselves. Their cocoons will be visible, and curious hands won't be able to touch them.

Once the chrysalides have formed, no further care is needed, just patience. In two weeks, butterflies will begin emerging. A day or two afterward, the children can watch as you release the butterflies outside.

From caterpillars to butterflies, the monarchs will be in your science corner for three to five weeks, depending upon how many caterpillars you capture and how large they are when you do. During this time, there are any number of science skills your students can learn from them.

Observing

Ask the children to observe the caterpillars. How big are they? What color are they? What are their bodies shaped like? Then give the children an outline of a caterpillar for them to color and stripe as if it were a monarch caterpillar.

Have the children carefully watch the caterpillars that are eating the milkweed leaves. Ask them, How fast do the caterpillars eat? Do they rest in-between bites? Then, take the class to a field with milkweed plants and have them observe the eaten leaves.

Next, the class should study the caterpillars at rest. Are the caterpillars still eating? What are they eating? Caterpillars eat their old skin as they shed it.

When the caterpillars are about 3 cm. long, they are fully grown. Soon thereafter, they start spinning a small button of thin white threads. Just before this happens, the caterpillars' stripes take on a greenish cast. Have the children watch the caterpillars closely. What does the caterpillar do with the button? It turns around and fastens its back pseudo-legs firmly on either side of the button, hanging from it in a J shape.

As the caterpillars split their skin and change into chrysalides, have the children observe the animals every two hours for about a day. A caterpillar will hang in the J shape for eight to 12 hours before changing into a chrysalis. Ask the children to color a picture of a chrysalis with the correct markings on a provided outline. On a daily basis, have the children check the chrysalides. In nine to 15 days, the chrysalides should appear darker, then turn black with orange markings. A day later, the monarch butterflies will appear.

As a butterfly emerges, have the children observe the size of the butterfly and the shape of its wings. Then have the children observe it a half hour later, then two hours later, four hours later, and again the next day.

When the butterflies are ready to fly, have the children note any special markings on the back wings. Black dots denote male scent glands.

Measuring, Charting, Mapping

You can measure the length of each caterpillar from the first day, using a piece of string. Construct a chart to show each caterpillar's growth.

Draw a chart of one-cm. squares. Have the children color a square for each day the caterpillar is in the chrysalis.

Using that chart, note how long the chrysalis lasted before the first butterfly emerged. Have the children predict how long the second chrysalis will hang before changing. Do this for each cocoon.

Once the butterflies have been released, show the children a map of Mexico. Talk to them about the monarch's annual migration and its life cycle.

Now that the children have observed closely the monarch's development, give them diagrams of the four developmental stages of the butterfly—egg, larva (caterpillar), pupa (chrysalis), adult butterfly. Have them color the diagrams and paste them in order on a colored sheet of paper. When they've finished coloring, they can fold their paper and make a booklet of their own monarch's development.

Finishing Up

These activities require children to make observations over many days. It helps them realize, too, that they will not always be there to see every step of the monarch's development. Possibly none of the critical developmental changes will occur while they are watching, yet the children can see the results. If they miss the butterflies emerging, show them a filmstrip on it. Once, however, our class turned off the lights to watch a filmstrip about butterflies coming out of their cocoons, and when we turned the lights back on, we had missed seeing our own butterflies emerge.

In our class, the butterfly unit usually encourages children to bring in other caterpillars they want to identify. Children might find dead monarchs and show them to you. If that happens, examine the wings and antennae with a microscope or magnifying glass. The children will be surprised at the scales on the wings.

It's not too early to think about bringing the wondrous monarch to next year's class. It's quite possible that if you don't think about them now, you might miss them altogether this fall.

Resources

Carle, E. (1969). *The very hungry caterpillar*. Cleveland: Collins World. [This is a description of what a caterpillar ate in the week before it became a butterfly.]

Friskey, M. (1961). *Johnny and the monarch*. Chicago: Children's Press. [This is the story of a boy who chases a butterfly.]

Mitchell, R., and Zim, H. (1964). *Butterflies and moths*. New York: Golden Press. [This is a field guide to butterflies and moths.]

Sabin, L. (1982). *Amazing world of butterflies and moths*. Mahwah, NJ: Troll Associates. [This is a book about the life cycles of butterflies and moths.]

Urquhart, F. (1976, August). Found at last, the monarch's winter home. *National Geographic Magazine*, pp. 161–173.

Madalyn Brown teaches first grade at Schoharie (NY) Elementary School. Jeffrey R. Lehman is an assistant professor of science education at State University of New York at Albany. Artwork by Max-Karl Winkler.

Phyllis R. Marcuccio

The monarch caterpillar (left) and chrysalis (right).

Phyllis R. Marcuccio

Tiger Salamanders

David C. Kramer

Tiger salamanders are widespread in the United States, from Canada south to the Rio Grande and the Gulf of Mexico and—with the exception of the Appalachian highlands

---

If progress unsettles a salamander in your neighborhood, help it relocate—or take it to school.

---

and the Northeastern states—from the Rocky Mountains east to the Atlantic Coast. They are the largest of the terrestrial salamanders, and the heavy-bodied adults reach a length of 17–23 centimeters (cm). About half of the tiger

Reprinted from *Science and Children*, David C. Kramer, February 1985, pp. 22–23

salamander's length consists of the laterally flattened tail. The head is broad and flattened with widely set eyes and a wide mouth containing teeth that are numerous but small and harmless. The tiger salamander's ground color is usually dark brown to black, and the body is generously covered with yellow to olive-brown spots of irregular size and shape.

Tiger salamanders, like most other amphibians, are sometimes said to lead double lives because the juveniles are primarily aquatic and the adults are terrestrial. During the summer months, the adults live in woodlands, forests, prairies, and open fields. Here they spend most of their time burrowing in the leaf litter and soil, where they find their natural diet of insects, insect larvae, and earthworms. Because of their burrowing nature and secretive habits, they are rarely seen during the summer and are sometimes referred to as "mole salamanders." However, these salamanders are commonly seen in the early spring, when the adults migrate to marshes, ponds, and prairie potholes to

mate and lay eggs before returning to their terrestrial way of life. The eggs hatch within a few days into aquatic juveniles, which, except for the presence of gills, look much like tiny adults. These larvae feed voraciously on water fleas (daphnia) and other similar aquatic organisms and grow to 7–10 cm before losing their gills and entering into the adult stage. This growth and transformation usually requires one summer but in the northern part of the salamander's range may require two seasons.

Although the major migration to and from breeding pools occurs in early spring, these salamanders are sometimes stimulated to move about by prolonged periods of rainfall and are often encountered at such times. During their periods of activity, and especially in the spring, tiger salamanders are often found in basements, garages, or window wells. This is especially true when new houses have been built in the traditional migration route of the salamander or near a breeding pool. In this case, a salamander may wander in while attempting to make its way around the newly constructed obstacle. If this should happen, the salamander can simply be removed and released a meter or so from the house (or taken to school, of course). During the winter months, tiger salamanders hibernate in underground tunnels or burrows made by other animals.

Housing and Care

The eggs, larvae, or adults can be kept in the classroom. A five-gallon (40-liter [l]) aquarium will be a satisfactory container for any stage, but the eggs and larvae can also be maintained and observed easily in a one-gallon (4-l) jar.

Tiger salamander eggs are about 3 millimeters (mm) in diameter, and each is contained in a clear gelatinous sphere that is 1.25 cm in diameter. There are 6 to 20 (and sometimes even more) eggs attached to a weed or twig in the pool. If the eggs are collected, they should be kept in water also collected from the pond with no more than 8 to 10 eggs to each 4 l of water. This will help ensure that the developing embryos have enough oxygen. The eggs will hatch in 8 to 10 days.

Salamander larvae, whether hatched from eggs in the classroom or collected directly from a pool, should be fed daily with daphnia or other similar aquatic organisms collected with an aquarium dip net. Since collecting live food every day or two is such a chore, salamander larvae are difficult to keep over a long period of time, but they are exceptionally interesting to observe on a short-term basis.

Adult salamanders tend to bury themselves if kept in a terrarium and are thus difficult to observe, but they can be kept in an aquarium with 2.5-5 cm of water for one or two weeks. They do not need any basking sites or other special appointments, and their food, consisting of mealworms, insects, or earthworms, can be placed directly in the water. Tiger salamanders will quickly learn to accept food held in forceps, and once they have learned this, they can also be fed strips of lean raw meat and fish. Uneaten food should be removed and the water changed regularly because tiger salamanders, like other aquatic animals, can absorb toxic wastes through their skins if these wastes are allowed to accumulate.

Like most other ectothermic organisms, adult salamanders do not have to eat every day, and if otherwise in good condition, they can pass the weekend without food. However, they are particularly susceptible to dryness and can dehydrate and die in a short time if they are allowed to run out of water. Finally, salamanders, like other classroom animals, should be returned to their natural environment after one or two weeks.

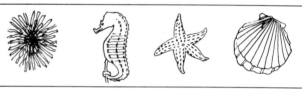

ARTWORK BY RICHARD GUY

Preserve Marine Specimens

When collecting marine specimens from coastal waters, place organisms in a plastic bag filled with seawater, wrap in newspaper, and pack in crushed ice in an insulated chest. (Salt in the seawater will prevent it from freezing.) Specimens will survive for about eight hours. BARBARA WATERS, Associate 4-H Agent, Cape Cod Extension Service, Barnstable, Massachusetts.

A Garden on a Button

Create a mini-forest with materials gathered on a nature walk and common household supplies.

Materials: rocks, twigs, moss, sand, white glue, and a cross section of wood about 10-20 cm in diameter and 4-6 cm thick (try to obtain a branch from which cross sections can be cut so that there will be bark around the circumference).

Procedure: Brush top of wood button with glue. Sprinkle with sand. Dab sand with glue and secure pieces of moss (with soil intact). Add pieces of twigs, rocks, etc., to create the appearance of a mini-forest. Allow glue to dry for a week. Water the moss once a week with a spray container or mister. Idea presented at the Illinois Science Teachers Convention in Chicago by LILIAN WILLIAMS, East St. Louis, Illinois, and SUSAN LINKSVAYER, Highland, Illinois.

Reprinted from *Science and Children*, March 1978, p. 40.

Reprinted from *Science and Children*, David C. Kramer,
November/December 1987, pp. 42-43.

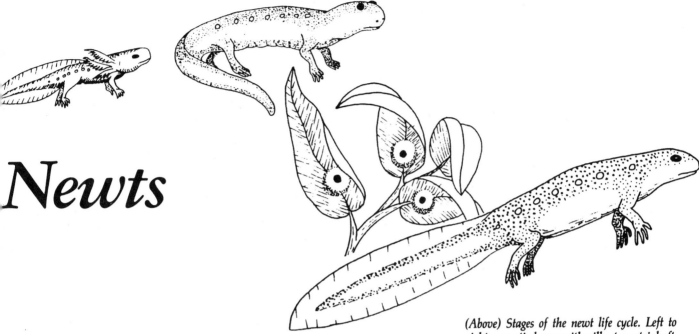

Newts

(Above) Stages of the newt life cycle. Left to right: aquatic larva with gills, terrestrial eft, eggs on aquatic vegetation, aquatic adult.

"What *is* a newt?" Although widespread and sometimes abundant, newts are not familiar to many people. Newts have a secretive nature, and when they do come out in the open, people think they're just more salamanders.

Indeed, newts are salamanders and resemble other woodland salamanders with long slender bodies, long tails, and about equal-length legs. Like most other salamanders, the newt life cycle has two distinct phases: aquatic and terrestrial.

Newts are also unique in several ways. Unlike other salamanders, their skin is rough and relatively dry; they lack distinct, vertical grooves on the body; the adults are largely aquatic; and one stage of their life cycle does not occur in any other amphibian.

From Egg to Eft

The typical salamander life cycle has three stages: egg, larva (or tadpole), and adult. Their eggs are laid in water, and hatch into aquatic larvae, which then become terrestrial adults. The adults return temporarily to the water to deposit their eggs.

In contrast, the newt life cycle has four stages: egg, larva, juvenile, and adult (see figure). Like other salamanders, newts deposit their eggs in water, where the larvae hatch, but here the similarity ends. Newt larvae transform into terrestrial juveniles called efts—a stage only newts exhibit. The efts eventually transform into permanently aquatic adults—again, unusual for salamanders.

When the efts first transform from the larval stages they are about 3 cm long. Efts of the red-spotted newt (the most widespread of the various species) are orange to brick-red in color and have small black circles with red centers on their sides. Efts of other species are generally some shade of brown and free of markings.

The secretive efts hide under fallen logs, pieces of bark, rocks, or leaf litter on the forest floor, sometimes far from their home pool. Rainfall stimulates them to move about, and they sometimes appear in great numbers on the forest floor during warm, rainy nights. They'll feed on worms, insects, and spiders as they grow slowly—and remain in the eft stage—for two to three years. When they grow to about 8 cm long, the efts become adults.

Aquatic Adults

Transformation into the aquatic adult stage involves several changes. The newt's body becomes more streamlined and a fin develops both above and below the tail. The coloring also changes. In the case of the red-spotted newt, the adults are olive-green to yellow, retaining the black rings with red centers. In other species, adult coloring tends to be a lighter shade of the eft coloring.

The adult stage also brings changes in the circulatory and digestive systems necessary for an aquatic existence. Of course, the newts also reach sexual maturity at this time.

Adult newts most often live among the vegetation in ponds, the shallows of lakes, or the backwaters of rivers. Typically feeding on a variety of aquatic insects, small snails, leeches, and worms, adult newts have also been known to eat frog eggs and tadpoles. They often rest on the bottom of the pool or drift among the vegetation where their coloring protects them from predators such as fish and certain wading birds.

Most amphibians must return to the water to reproduce. Since adult newts live in water continuously, their return to water occurs when the efts transform into adults. Courtship may take place in the fall, winter, or spring. Some time later, the female deposits hundreds of gelatin-covered eggs, attaching them to the aquatic vegetation. The eggs usually hatch in 20 to 30 days. The

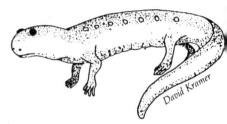

David Kramer

larvae, which look like tiny salamanders with external gills, feed on minute aquatic organisms. After about three months, when they reach about 3 cm in length, the larvae transform into efts to continue the life cycle.

Newts can be found in the Pacific and Atlantic coastal states, and from the east coast westward to just beyond the Mississippi River and extending into Canada.

Collecting Newts

To collect efts, search under logs, pieces of bark, or other objects on the forest floor. (Take care to return each object to its original position to protect the micro-habitat.) Alternatively, efts may be found at night; use a flashlight to spot them during or following a warm, summer rain. To find the adults, carefully search through the shoreline vegetation of a quiet, shallow lake or

---

In The Classroom Animal, a development of S&C's popular Care and Maintenance series, column writer David C. Kramer focuses on the natural history of small animals suitable for short-term classroom study and on how to care for these animals. Readers wishing to communicate with Professor Kramer should write him at the Department of Biology, St. Cloud University, St. Cloud, Minnesota 56301.

---

pond. The adults move faster than the efts, but they, too, can be caught by hand. It will be better for the newt, though, if you use a long-handled dip net. Also, some pet stores sell adult newts, usually calling them "water newts."

Housing and Care

The housing requirements for newts, of course, depend on their life cycle stage. Efts need a forest floor terrarium, preferably constructed from materials found where the efts were collected. Place 3 to 6 cm of moist forest soil in the bottom of a 10-gallon aquarium to create a moist substrate for the efts and a rooting medium for some small native plants or ivy. A handful of dried leaves, some twigs, and a piece of bark arranged on the soil surface make the habitat look natural. (Efts do not require soil or plants, but proper care of the plants—moist but not wet soil and indirect light—will help assure a favorable environment for the efts.) A variety of small, forest floor organisms such as worms, isopods, and insects should also be added as food for the efts. (See "Cryptozoa," *S&C, 24*(6):34–36.) While the efts cannot climb out of an aquarium, a glass cover should be used to help maintain humidity. Add a small dish of water to supplement their moisture from the air and soil.

Adult newts must be kept in water. An aquarium with a sand or gravel bottom and some water plants is an appropriate environment for them. Either aquatic plants collected from the newt's natural habitat or those from a pet shop will be suitable. A heater is not needed, but the plants, of course, require some indirect natural light or an artificial light source. Adult newts will eat small worms or insects and can learn to take small strips of red meat or fish from forceps. Two or three times each week, feed them the amount of food they can consume in a few minutes. Remove any uneaten food from the aquarium to prevent it from spoiling.

Finally, when the project is over, the newt, whether eft or adult, should be returned to its natural environment.

Resources

Mattison, Christopher. (1982). *The care of reptiles and amphibians in captivity.* Dorset, U.K.: Blandford Press.

Palmer, Lawrence E., and Fowler, H. Seymour. (1978). *Fieldbook of natural history* (2nd ed.) New York: McGraw-Hill.

Simon, Seymour. (1975). *Pets in a jar: Collecting and caring for small wild animals.* New York: Viking.

Vallin, Jean. (1968). *The animal kingdom.* New York: Sterling.

Reprinted from *Science and Children*, David C. Kramer, April 1985, pp. 35-36.

Leopard Frogs

Leopard frogs occur from southern Canada to Mexico and from the Atlantic coast westward almost to the Pacific Ocean. They are the most widespread and probably the best known of all North American amphibians. Thus, the leopard frog is likely to be the image that comes to most people's minds at the mention of the word *frog*. Leopard frogs serve as fish bait, and they have been widely used for dissection in biology classes. Five to eight centimeters (cm) long as adults, leopard frogs may be various shades of brown or green, and they have black spots about the size of a pea distributed irregularly over the back and sides. Their bellies and the undersides of their hind legs are white.

Habits and Life Cycle

The preferred habitat of leopard frogs includes practically any type of fresh water from weedy shorelines of large lakes to small marshes and streams. However, these frogs often venture far from water, and this tendency probably explains the other names by which they are known—*grass frogs* or *meadow frogs*.

In the autumn leopard frogs migrate to the bodies of water that will be their breeding pools the following spring. Entering the water, they bury themselves in the bottom mud, where they hibernate throughout the winter. Mating occurs when the frogs emerge from hibernation; and the females produce hundreds of gelatin-covered eggs, which form a mass from 10–15 cm in diameter. The eggs hatch in a few days into tadpoles, which feed on algae and decaying vegetation for two to three months

John C. Coulter

(depending mostly on water temperature and available food), until they are transformed into adults.

This transformation is an amazing feature of their life cycle. First the hind legs appear; then the front legs. The mouth changes from a small oval to a wide slit, and the tail is absorbed. The external changes are accompanied by similarly spectacular internal changes as the tadpoles metamorphose from herbivorous aquatic animals to air-breathing, terrestrial, insectivorous ones. At the time of their transformation, leopard frogs are 4–5 cm in length.

Although leopard frogs are prolific, producing thousands of eggs in each mating, both the tadpoles and adults are preyed upon, and their numbers have been held in balance by their many predators—fish, reptiles, birds, and mammals. In recent years, however, for reasons that are not entirely clear, the population levels of these frogs have declined drastically in many areas. This trend now seems to be reversing itself in some areas where numbers are increasing and leopard frogs are approaching former population levels.

The natural foods of leopard frogs—insects, spiders, caterpillars, and occasionally earthworms—are caught on a frog's moist tongue, then quickly drawn into the mouth and swallowed.

Housing and Care

A 20- or 40-liter aquarium is an ideal container for captive leopard frogs, but it must have a cover to prevent specimens from jumping out, and it must be set up to allow the access to water, dry land, and a hiding place. Like other amphibians, leopard frogs have moist skins that permit the loss or absorption of water, so they must always have 2.5–5 cm of fresh water available, but they should also be able to leave the water. Such an arrangement can be provided in several ways.

Many sourcebooks recommend building up one end of an aquarium as a land habitat with soil or gravel and leaving the other end as a water habitat, but this arrangement makes cleaning the cage and changing the water difficult. Simply placing a brick or one or two large rocks in the aquarium with 2–5 cm of water achieves the same effect and is a more practical method for creating temporary quarters. However, it does not supply the third important need for the habitat, a hiding place. The natural reaction of frogs is to leap when alarmed, and a captive is likely to jump against the glass and injure itself unless it has somewhere to hide. An environment that is safe, healthy, attractive—and practical—can be set up in a planted terrarium (5 cm of soil with several bushy plants) that also contains a water dish 5 cm deep. Or, a gravel substrate can be used, with a potted plant and a piece of bark or wood to create a hiding place.

To feed a frog, put live insects, mealworms, or earthworms in the cage near it. Leopard frogs will also accept bits of lean meat dangled in front of them on a string.

A Note on Health

Captive frogs sometimes develop a disease known as red-leg. However, this is not likely to occur if they have the opportunity to climb out of the water, if they have plenty of room (avoid crowding several frogs into the same enclosure), and if fresh water is provided daily. Also, frogs usually eliminate waste into the water as they are sitting in it—another important reason for providing fresh water daily.

> **Suggested Activities, Observations, and Questions**
> • Observe and describe how frogs catch and swallow their prey.
> • Place a frog on the floor and measure how far it can leap.

Collecting and Maintaining Frog and Toad Eggs

Frogs and toads lay eggs from early March to midsummer in still ponds and flowing streams. The time of egg laying, size of each egg, and egg mass varies with species. Some kinds of frogs lay their eggs on floating or submerged aquatic vegetation; others lay their eggs along the edge of the water. In streams, look for eggs in the quieter eddy pools.

To collect eggs, place a plastic bucket beneath the egg mass. If the eggs are on aquatic vegetation, cut the vegetation above and below the egg mass so as not to disturb it. If the eggs are unattached, place the bucket adjacent to the egg mass and allow the eggs to be pulled into the bucket. Toad eggs occur in long chains and bullfrog egg masses are usually large enough to fill 2 one-gallon containers.

Eggs should be brought to the classroom in large plastic buckets, about two-thirds full of water. A pitfall to keeping frog eggs successfully is placing too many eggs in too small a container where the oxygen supply is limited, both in transport to the classroom and also in rearing.

Maintaining Eggs

Frog eggs are best maintained in spring water or previously boiled pond water. Place no more than five frog eggs in each quart container full of water. Each small group of students will have their own eggs if you set up several jars in this way. To separate frog eggs from a large mass, cut through the clear jelly-like mass with a single-
Continued on page 74

Reprinted from March 1974, *Science and Children*, p. 36. This article was the first in the "Care and Maintenance" series and was prepared by Paul Hummer.

Reprinted from *Science and Children*, David C. Kramer, October 1985, pp. 30-31.

Tree Frogs

There are several kinds of treefrogs in North America. Most of them are found only in the southeastern region of the United States, but the two most widely distributed kinds, the gray treefrog and the spring peeper, are found throughout the eastern half of the country and into southern Canada. Three other treefrog species occur in the southwestern states, and the range of one of these, the Pacific treefrog, extends northward along the West Coast into Canada.

Characteristics

Treefrogs are relatively small compared to most other frogs. Depending on the species, full grown adults usually range from 2–5 cm in length. Although occasional specimens may be robust looking, treefrogs are generally slender-bodied, with long, slender legs. Their skin is slightly warty, and as a result they are sometimes called "tree toads." Treefrog coloration is typically green or brown, and some species have dark markings on their backs. The most important distinguishing characteristic of treefrogs, however, is that each toe is tipped with a disk-shaped pad that helps them cling to surfaces as they climb about.

The typical habitat of treefrogs is damp forests and woodlands. Being nocturnal, treefrogs spend the daylight hours either hidden among the litter on the forest floor or sitting quietly on a limb or leaf. At night they climb silently through the trees and shrubs in search of their insect and spider prey. Because of their secretive habits and the camouflage provided by their coloration, treefrogs are not often noticed by human beings. They are occasionally seen, however, when their nocturnal search for food brings them to a window or porch to catch insects that have been attracted to the light. Here, they sometimes give a striking demonstration of their climb-

David C. Kramer

In search of insects, a treefrog climbs a window. The disk-shaped pads on its toes help it cling to and climb on the glass.

ing ability as they climb up windows in pursuit of insects. Otherwise, their presence can be detected when they call, sometimes persistently and in large numbers, from a spring breeding pool or, later in the summer, from an isolated perch in a tree.

The reproductive habits of treefrogs are similar to those of other frogs and toads. Soon after emerging from hibernation in the spring, male treefrogs enter shallow pools and begin calling. The females respond to the sound and enter the pools, where mating and egg laying occurs. The adults then leave the water to live among the trees and shrubs for the remainder of the summer. The eggs soon hatch and the diminutive tadpoles feed on algae and decaying vegetation for several weeks before entering the adult stage. Thereafter, the young frogs leave the water to lead a life similar to that of the adults until the time comes, in the autumn, for hibernation.

Treefrogs are among the easiest and

most interesting of all amphibians to keep in captivity. They are docile and easily handled. Although they can jump short distances, they are primarily climbers and are not inclined to make the sudden and surprising leaps characteristic of other frogs. They will sit quietly for hours; yet they are alert and quickly become active predators when presented with their live insect food. And, once provided with appropriate enclosure and food, treefrogs require little attention and care.

Housing and Care

A well-established terrarium, especially one that simulates the treefrog's natural habitat, is an excellent place to keep a treefrog. The plants make good climbing and hiding places, and the moisture in the soil and that given off by the plants helps maintain an adequate level

David C. Kramer

of humidity. A little moss, a few dried leaves, and perhaps a rock, a piece of bark, and a stick or two will give a more natural appearance and add additional climbing and hiding places. Of course, a cover is necessary to prevent the frog from escaping. While a screen cover is sufficient, a glass one is preferable as it will help maintain the high humidity important to the frog and the plants.

A convenient alternative to a planted terrarium, and one that has most of its advantages—except for aesthetic appeal—can be created by placing a potted plant in a closed container large enough to accommodate it. An even less sophisticated arrangement—but a satisfactory one on a short term basis—can be provided by putting a few moist paper towels in the bottom of the container and adding a stick or two for climbing. Since treefrogs are relatively inactive, the size of the enclosure is not crucial as long as it is not too confining. A standard aquarium is more than adequate, and a plastic sweater box or 3-L jar is sufficient.

Whatever the arrangement, remember that treefrogs, like most other amphibians, are susceptible to dehydration, and maintaining a high humidity level in the cage is necessary for their comfort and survival. Moist soil and plants will sustain the necessary moisture level. But, even though treefrogs rarely, if ever, drink, they should also have a container of water in which they can soak to maintain their water balance if necessary.

The normal food of treefrogs is live insects and spiders, which they catch in their arboreal environment. In captivity, they should be provided with the same fare. A sufficient quantity of insects can usually be collected at night around a lighted window or during the day under rocks and boards. Alternatively, treefrogs will eat mealworms or crickets that can either be raised or purchased from a local pet store. Of course, the food, whether insects or spiders, must be alive as it is movement that stimulates the treefrog's feeding reaction. It is surprising how large an object these frogs can consume in proportion to their size, but they seem to prefer food items that are approximately one-fourth to one-third their own length.

Although the temptation to keep a specimen for an extended period is great, it is difficult to provide live food for a long period of time. For this and other reasons, it is best to keep a treefrog for only a few days and then release it into its natural environment.

Resources

Conant, Roger. (1975). *A guide to reptiles and amphibians of eastern and central North America*. Boston: Houghton Mifflin.

Mattison, Christopher. (1982). *The care of reptiles and amphibians in captivity*. Dorset, U. K.: Blandford Press.

Palmer, Lawrence E., and Fowler, Seymour. (1975). *Fieldbook of natural history*. (2nd ed.). New York: McGraw-Hill.

Smith, Hobart M. *Amphibians of North America*. New York: Golden Press.

Continued from page 72

edged razor blade or scissors. Fertile eggs will be dark on top and lighter on the bottom. The developing embryo floats to the top of the egg. Infertile eggs will not develop this characteristic color separation. They should be removed. Some eggs will have cotton-like fibers attached to them. These are aquatic fungi that attack dead or infertile eggs. If possible, remove these eggs.

Keep the jars containing the eggs between 62° and 72°F (17°–22°C). Changes in the developing eggs can be observed with a small hand lens or stereomicroscope. Eggs to be observed may be transferred to small plastic cups by using an aquarium net or plastic spoon.

No food is necessary for the developing egg until the yolk disappears. Many varieties of food can be fed to small tadpoles; however, my experience has been that small amounts of goldfish food are satisfactory. Do not overfeed, as this will foul the water. Add a small amount of food and watch the tadpoles clear the container of food. Overfeeding is as bad as underfeeding. Change the water at least two times a week or when it becomes very cloudy.

Housekeeping Supplies

A plastic household baster is a handy tool for removing debris from the bottom and adding oxygen to the jar. To add oxygen if an aquarium aerator is not available, fill the baster with water from the jar containing the eggs or tadpoles. Hold the baster three to four inches above the jar and expel the water into the jar. Repeat this 8-10 times. Note the bubbles being forced into the water. This procedure should be repeated each day.

Jars containing eggs or small tadpoles should not be exposed to direct sunlight. Keep the jar in a cool part of your room.

Hatching Snake and Turtle Eggs

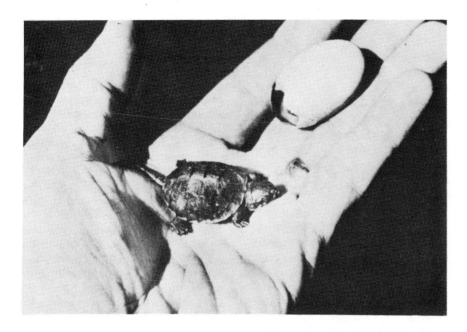

When a snake or turtle being kept as a classroom pet lays eggs, the occasion generates considerable enthusiasm. This is not an uncommon event, particularly if the pet is a turtle picked up in the spring as it was wandering about prior to nesting. It is interesting to observe reptile eggs and compare their size, shape, and texture to the more familiar eggs of birds. Even more interesting is the possibility of hatching the eggs.

Hatching most snake or turtle eggs is a simple matter and may be accomplished in several ways. The method described in this article has the double advantage of not requiring special equipment such as an incubator, and once started, requiring little attention. The only materials needed are a thin plastic bag and some paper towels.

Moisten two or three paper towels, squeeze them out as thoroughly as possible, then form them into a cup-shaped nest just large enough to accomodate the eggs when arranged in a single layer. Place the paper nest and eggs in the plastic bag. Close the bag and secure with a wire fastener placed as high as possible on the bag so that an air-space is present above the eggs.

Important Considerations for Hatching

Now, four factors including temperature, moisture, mold prevention, and care of the hatchlings should be considered. First, a precise temperature is not critical for hatching most reptile eggs. Temperatures from 22° to 33° C (70° to 90° F) are acceptable; room temperature is therefore satisfactory. The important consideration, as far as temperature is concerned, is that *the eggs should not be chilled or overheated either before or during the incubation process, and the temperature should not be allowed to change too suddenly.*

It is important to keep the eggs moist to prevent dehydrating but *too much water should be avoided.* A good rule of thumb is to keep the eggs only slightly moist but if they start to dehydrate, as evidenced by a slight indentation of the flexible shell, sprinkle a small amount of water on the towels.

Whenever an organic material is exposed to high humidity and moderate temperatures it is susceptible to molds. Reptile eggs are no exception. The best way to help prevent mold growth is to keep the eggs as dry as possible without allowing them to dehydrate. However, if mold does develop, it should be carefully scraped off of the eggs and the paper towels should be replaced. This may have to be done three or four times during incubation.

The final consideration, that of caring for the young, is also a simple matter. During the final stages of development, reptiles absorb a large amount of yolk which sustains them for a long time. Therefore, the young will not require food for a week or two. At this time they should either be released in a suitable habitat or the proper food should be provided. *Water should be provided for the animals at all times.*

Don't Give Up!

One should not expect all of the eggs to hatch. Some might not be fertile. Also, reptile eggs require a long time to hatch—in some cases, three months. Even after the eggs begin to open, several days may pass before the young emerge. At any rate, the interest produced by this activity will be well worth the wait!*

Check with local government offices or wildlife specialists for facts about the protected species in your state. (You may have a hot turtle!)

Reprinted from *Science and Children*, David C. Kramer, March 1974, p. 31.

John Coulter

Painted Turtles

ainted turtles are probably the best known of all North American turtles. Though they are rare or absent in the southern tier of states, they are the most abundant of all turtles in the northern states and southern regions of the Canadian provinces. The carapace, or upper shell, of painted turtles is generally smooth and, except for an intricate lacelike pattern of red, yellow, or black along the margin, is drab olive to gray. The head, neck, legs, feet, and tail are black with bold yellow or reddish yellow stripes. The plastron, or lower shell, also a yellow to reddish yellow, is plain in eastern specimens but has a dark central blotch in the western forms. Painted turtles range in size from about 2.5 centimeters (cm) long at hatching to 15–17 cm long as adults.

Behavior

Although painted turtles occasionally wander about on land, they are primarily aquatic and seem to prefer quiet, shallow water with abundant submerged vegetation: marshes, swamps,

ponds, roadside ditches, river backwaters, and even the shallow margins of larger lakes can all support large populations of painted turtles. Painted turtle food consists largely of the aquatic vegetation in which turtles hide and the abundant insects and snails that live in this vegetation. These turtles spend part of every day basking in the Sun, and they bask by floating at the surface of the water if there is no surface for them to climb on. If painted turtles are disturbed while basking, they will slide into the water and hide themselves. Otherwise, they are curious and will sometimes swim close to a boat or dock to observe any activity there.

Painted turtles are active during the warm summer months, but long before the water freezes, they will bury themselves in the muddy bottom of their home pool, where they hibernate until late spring. Mating occurs soon after the turtles emerge from hibernation, and the females then journey onto land, where they bury their clutch of six to ten eggs in the soil. The eggs normally hatch after six to eight weeks, and the young make their way to the water where they, like the adults, feed on aquatic vegetation, insects, and snails.

Housing and Care

Since painted turtles are primarily aquatic, a classroom turtle should be housed in a container in which it can submerge itself completely. But it should also be able to leave the water at will. A

40-liter aquarium with 7–12 cm of water and a rock (or a brick, inverted flowerpot, or water bowl) that extends above the waterline is a simple arrangement that meets the turtle's needs and is easy to maintain. Some sourcebooks suggest a gravel bottom that slopes so that half is submerged and half is not. However, this terrarium/aquarium arrangement, while functional, is less desirable than the arrangement mentioned above because the gravel tends to become fouled with food particles and droppings and is difficult to clean. Since turtles are messy eaters, the ease with which the enclosure can be cleaned is certainly an important consideration.

Painted turtles do well in a normal classroom environment (even if the temperature is reduced over weekends). However, they seem to like warmth, and they will thrive in a warm part of the classroom. If there is reason to believe that cool temperatures are a problem, place a lamp above the basking rock.

Diet and Feeding

In captivity, painted turtles prefer to eat their normal diet, which includes certain aquatic vegetation, insects, and snails. These items are sometimes difficult to provide, so mealworms, earthworms, and small strips of fish or lean meat can be used as substitutes for a short time (two or three weeks). The food can simply be put in the water, but any food not consumed within an hour or two should be removed to prevent its decaying and fouling the environment. After two or three feedings, one can easily estimate the proper amount of food to provide. These turtles will also quickly learn to take food from a forceps, and this is a good feeding technique as, once the turtles have learned it, they will demonstrate when they've had enough to eat by losing interest in the food being offered. Normally,

Reprinted from *Science and Children*, David C. Kramer, May 1985, pp. 42–43.

painted turtles should be offered some food each day, but these turtles can go two or three days without feeding, so it is easy to take care of the weekend with a good feeding on Fridays.

Long-term Captivity

Turtles are an excellent example of animals that are easily kept for two or three weeks but present greater problems if kept for extended periods. Though the foods mentioned above will sustain them adequately for a short time, they might not satisfy the turtles' long-term nutritional needs—for example, calcium, which is lacking in mealworms, and vitamin D, which they normally produce by basking. Therefore, if painted turtles are kept for more than one month, their diet should be fortified with vitamin and mineral supplements. Furthermore, some specimens may become lethargic during the winter months, when they normally hibernate. At this time, they might refuse to eat, and, if so, their health might decline and they would become a burden rather than a valuable classroom resource. It is probably easiest for the keeper and best for the captive to release a painted turtle into its natural environment after a period of two or three weeks.

A Note on Disease

Turtles are known to be carriers of salmonella bacteria, the cause of salmonellosis in humans. According to Christopher Mattison, "The risk is not very great, provided that normal hygienic measures (washing their hands) are taken after handling the animals or cleaning their quarters." Since the bacteria are eliminated with the feces, removal of the droppings and frequent changing of the water is important. Finally, although the danger of disease is not great, it is probably most prudent to have students observe but not handle captive turtles.

Resources

Conant, Roger. *A Field Guide to Reptiles and Amphibians of Eastern and Central North America.* Boston: Houghton Mifflin, 1975.

Ernst, Carl H., and Roger W. Barbour. *Turtles of the United States.* Lexington: University of Kentucky, 1973.

Mattison, Christopher. *The Care of Reptiles and Amphibians in Captivity.* Dorset, U.K.: Blandford Press, 1982.

Suggested Observations, Activities, and Questions
- Watch the turtle eat. How does it bite and swallow its food?
- How long can the turtle remain submerged?
- Observe and describe how the turtle swims.
- Why do turtles bask?

John Coulter

Reprinted from *Science and Children*, David C. Kramer, February 1986, pp. 55-57.

Box Turtles

There are several kinds of box turtles in the United States. They differ only slightly in appearance, but because they occupy different habitats, they have different environmental needs and food preferences. Box turtles are found throughout the eastern and central United States, from southern Maine westward to South Dakota, then southward to Arizona. (See map.)

Appearance

The box turtle is renowned, and appropriately named, for its ability to pull in its head, feet, and tail and close its shell for protection. This feat can be accomplished because the turtle's high, dome-shaped *carapace* (upper shell) provides room for its appendages, and because the *plastron* (lower shell) is hinged, allowing the two halves to be pulled upward and held tightly closed. This shell makes the box turtle the best protected of all North American turtles.

In turtles, there is an apparent relationship between the degree of protection afforded by the shell and the relative aggressiveness of the species. The snapping turtle, for example, has a reduced plastron and consequently less protection from its shell, but this is more than made up for by its aggressiveness. Conversely, the complete protection afforded the box turtle by its shell is reflected in its retiring and docile nature.

The box turtle, like the giant tortoise it resembles, is adapted for a terrestrial rather than an aquatic envorinment. The shape of the carapace and the degree of protection it provides, the short, stumpy legs and small feet, and the lack of webbing between the toes are all terrestrial adaptations. The toes are equipped with strong claws for digging in the soil.

The box turtle's shell is usually black or brown, overlaid with yellow to orange spots or streaks. However, the mark-

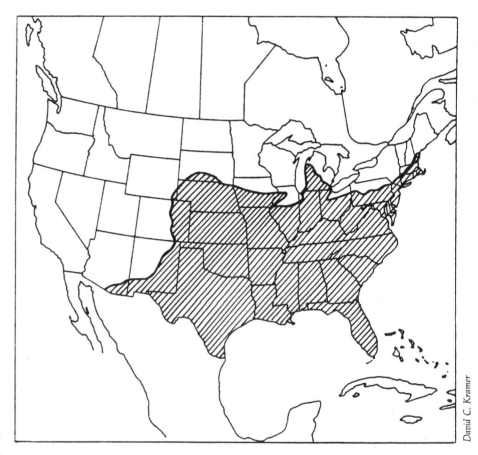

In the U.S., box turtles live in the areas shaded.

David C. Kramer

ings are highly variable. Occasional specimens display few or no markings, while in others the lighter colors are so profuse that the shell appears to be yellow or orange with irregular dark markings. Markings on the head, neck, and legs are equally variable.

The eyes of the young box turtle are normally brown, but in mature males they are often red. Females rarely have red eyes. This fact is useful in determining the gender of a specimen, but by itself it is not totally reliable. Another sexual difference, although also variable, is that the rear lobe of the plastron is slightly concave in males to allow closer contact during mating. These two characteristics taken together generally make it possible to determine gender.

Habitat

River bottoms, moist forests, and woodlands are typical habitats for the Eastern box turtle. Western species are more tolerant of dry conditions and more likely to be found in open areas. Box turtles are active throughout the warm months but are most conspicuous during the spring and fall as they wander about in search of food. In the summer, they avoid the midday heat by limiting their activity to the morning and evening hours. During extended dry periods, they bury themselves in moist soil or mud or soak in shallow water, sometimes for days at a time. In

John C. Coulter

disturbance, however, it retreats into its shell, and the entire procedure starts again. Most turtles lose this strong protective reaction after a short time in captivity and will withdraw only when strongly provoked.

The Classroom Turtle

Box turtles adjust nicely to captivity. Their short-term needs are simple and easy to supply. They are more active and alert than most other turtles and seem relatively more intelligent. For these reasons, they make interesting captives and are more commonly kept as pets than any other turtle.

Some turtles, however, are protected in some states. If there is any question about the legality of keeping a box turtle, check with a local conservation officer to be sure.

Box turtles, like other turtles (and most other vertebrate animals), can carry *Salmonella* bacteria, which causes salmonellosis in humans. Check out local health regulations before you decide to keep a turtle in school. And be sure to have students wash their hands carefully after handling or feeding the classroom turtle or cleaning its cage.

Housing and Care

A medium to large aquarium is an ideal enclosure because of the visibility it gives, but a wooden box or a child's plastic wading pool is also suitable. If the container is as much as three times the height of the turtle, no cover is needed. Otherwise, a screen cover should be provided. The interior appointments of the enclosure should include a water container larger than the turtle's body and a few centimeters of soil. The container will serve as a source of drinking water and also as a place for the turtle to soak itself. If the water container is more than 5 cm deep, put some rocks or gravel in its bottom to assist the turtle in climbing out. Then fill the outside enclosure with soil to about the top of the water container. The soil will provide a comfortable substrate and a medium in which the turtle can bury itself.

Diet and Feeding

Captive box turtles should be offered food resembling their natural diet. Most specimens will readily accept mealworms, crickets, and earthworms, all of which are easy to collect, raise in the classroom, or purchase locally. As a

the cool days of fall, they bury themselves in soft soil or enter the burrows of other animals where they hibernate until spring.

Reproduction

Box turtles typically mate in the spring soon after they emerge from hibernation. Then, in June or July, the female lays from three to eight leathery-shelled eggs in a nest she excavates in the soil. Under favorable circumstances, the eggs will hatch in two to three months, just in time for the hatchlings to enter hibernation. The young turtles usually go into hibernation without feeding, so they grow very little until the following year. Juvenile box turtles are so secretive that little is known of their early growth and activity. However, they appear to grow slowly, reaching sexual maturity and a length of 7.5 cm in about 5 years, 10 cm in 12 to 15 years, and 12.5 cm in about 20 years. Later growth is very slow, but box turtles continue to grow as long as they live. Box turtles commonly reach 60 years of age, and some may live for over 100 years.

Diet

The box turtle eats a wide variety of plant and animal materials, but it is an opportunistic feeder and generally consumes whatever is available. Unlike most aquatic turtles, the box turtle does not insist that its animal food be living.

Defense Reaction

When a box turtle is encountered in the wild, its first line of defense is to withdraw into and close its shell. Any attempt to encourage a turtle to open up by prodding will be futile. The only way to get one to come out is to leave it undisturbed for several minutes. It will then follow a predictable pattern of behavior and come out in its own time—turtle time, which is slow. First, the shell opens slightly and the turtle peeks out. If its surroundings look safe, it opens the shell wider and pokes out its head. After another careful look around, it opens still wider and slowly extends its feet. Finally, if there is no disturbance, the turtle will wander off and go about its normal activity.

Emerging from the shell might take a turtle 15 to 30 minutes. If there is any substitute for live animal foods, offer small (insect-size) strips of lean raw meat or hamburger. Some specimens will also learn to accept canned dog or cat food. To round out the diet, offer a variety of vegetables, including lettuce, and fruits, such as apple, cantaloupe, banana, and an occasional strawberry.

The food preference and the amounts consumed will vary, so do not be alarmed if some of these items are not accepted, or if the turtle does not seem to be hungry. A box turtle may not choose to eat at regular intervals, but infrequent meals will meet its needs as long as a sufficient quantity of food is kept available. The key to success in keeping a box turtle is to offer it a variety of foods and let it select its own diet. If a specimen has not eaten after two or three weeks, however, it should be released at the point of capture.

Long-Term Captivity

If a box turtle is being kept over the winter, it might bury itself in the soil and remain inactive for as long as eight weeks, if it is not disturbed. There are two reasons for this. First, the turtle's biological clock is probably telling it that it is time to hibernate. Second, the short days and cool nights (and weekends, if the heat is turned down in the school) tend to reinforce its biological clock. Since a classroom is not cool enough for hibernation through the entire winter, the turtle should not be allowed to remain inactive for more than a month.

To encourage the turtle to surface, suspend a light bulb near one end of the cage and adjust its height to produce a constant temperature of 25–30°C. The constant warmth and light will reduce the turtle's desire to hibernate and stimulate it to eat.

Since keeping a turtle over the winter offers these special problems, it is probably in the best interest of both keeper and captive to release the turtle before cold weather starts, either at the point of capture or in some other suitable habitat.

Resources

Conant, Roger. (1975). *A guide to reptiles and amphibians of Eastern and Central North America.* Boston: Houghton Mifflin.

Ernst, Carl H., and Barbour, Roger W. (1973). *Turtles of the United States.* Lexington: University of Kentucky Press.

Oliver, James A. (1955). *The natural history of North American amphibians and reptiles.* Princeton: Van Nostrand.

Discovering Your Friendly Neighborhood Bog

Not all schoolchildren are lucky enough to live near a famous swamp like Okefenokee. But everyone lives near "wetlands"—a marsh, a bog, a playa lake, an estuary, a fen—and we are beginning to value them rather than to destroy them uncritically (35 percent of U.S. wetlands had been drained and ruined by the 1970s). To help you find your neighborhood wetland and make it into a classroom rather than a parking lot, consult some of the resources below. Then take your students out to get their feet wet.

Although wetlands and their various subcategories defy precise definition (they may be flooded with salt or fresh water; they may be drained or under water at varying intervals), they share certain similarities:

• They are all saturated sometimes but not always (in which case they would be called "deepwater habitats").
• Their saturation determines the nature of their soil development and the types of animals and plants living there.
• This water creates severe problems for all flora and fauna except those specially adapted for life in it or in wet soil.

Call Aunt Samantha

Many government agencies cooperated with private sources to produce a colorfully illustrated booklet, *Status and Trends of Wetlands and Deepwater Habitats in the Coterminous United States, 1950s to 1970s.* Copies are available for $5 from the Department of Forest and Wood Sciences, Colorado State University, Fort Collins, CO 80523.

The Fish and Wildlife Service of the Department of the Interior has put together several useful "pacs" containing reproducible materials: instructors' overviews; two-sided colorful posters; puzzles or games; and three lesson plans or activities each. Those particularly helpful for studying wetlands are an issue pac, "Wetlands Conservation and Use," and a habitat pac, "Freshwater Marsh." Pacs are available for $5 (less for bulk orders) through the National Institute for Urban Wildlife, 10921 Trotting Ridge Way, Columbia, MD 21044.

The Council on Environmental Quality has coordinated an interagency report on *Our Nation's Wetlands,* which can be ordered for $5 from the Government Printing Office, Washington, DC 20402 (Stock Number 041-011-00045-9).

For Information on the Fen Nearby

Contact the Marine Education Materials System (MEMS) for access to its extensive database on wetlands material—curricula, pamphlets, articles, film guides, and activities. Write, enclosing $5, to Sue Gammisch, Virginia Institute of Marine Science, Gloucester Point, VA 23062, or call her at 804-642-7000. Anything you want to see on their printout can be ordered on microfiche for 75¢ or seen at one of the MEMS collections located nationwide.

In addition, you can write to the nearest regional office of the Fish and Wildlife Service (Attn: National Wildlife Refuges). A list of addresses for the regional offices.

• Lloyd 500 Bldg., Suite 1692, 500 NE Multnomah St., Portland, OR 97232; (For California, Idaho, Hawaii, Nevada, Oregon, and Washington)

• Box 1306, Albuquerque, NM 87103; (For Arizona, New Mexico, Oklahoma, and Texas)

• Federal Bldg., Fort Snelling, Twin Cities, MN 55111; (For Illinois, Indiana, Iowa, Michigan, Minnesota, Missouri, Ohio, and Wisconsin)

• 75 Spring St., SW, Atlanta, GA 30303; (For Arkansas, Alabama, Florida, Georgia, Kentucky, Louisiana, Mississippi, New York, North Carolina, South Carolina, Tennessee, and Puerto Rico)

• One Gateway Center, Suite 700, Newton Corner, MA 02158; (For Connecticut, Delaware, Massachusetts, Maryland, Maine, New Hampshire, New Jersey, Pennsylvania, Vermont, Virginia, and West Virginia)

• Box 25486, Denver Federal Center, Denver, CO 80225; (For Colorado, Kansas, Montana, Nebraska, North Dakota, South Dakota, Utah, and Wyoming)

• 1011 East Tudor Rd., Anchorage, AK 99503; (For Alaska)

Reprinted from *Science and Children,* May 1985, p. 8.

Green Anole (American Chameleon)

Previous articles in this series have dealt with animals that are widespread in the United States and can be easily captured by elementary school students. Green anoles differ in that they are found only in the southern states, from Texas eastward along the Gulf of Mexico and north through North Carolina. However, these small, attractive lizards are so readily available through pet stores and biological supply companies and appear so often in elementary classrooms as part of standard curricula or as classroom pets, that they deserve consideration here. This is especially true since, despite the best intentions of the keeper, many anoles die because inaccurate information about them leads to improper care. For example, at one time it was popular to recommend sugar water as adequate food for these insectivorous animals. Of course, anoles so fed soon died of malnutrition.

The adult male anole has a pink throat, which it can expand into a fan to signal other anoles.

As the erroneously applied common name "chameleon" suggests, these small, slender lizards, 12 to 20 centimeters (cm) long, are able to change their color. (True chameleons, the masters of color change, are native to Africa.) Green anoles are usually solid green on top, but the color can change to mottled green-brown or solid brown, depending on the specimen's mood and the environmental temperature. The lizard's chin and underparts are white (this does not change), and adult males have a pink throat, which can be expanded to signal other anoles.

In nature, anoles live among the twigs and leaves of shrubs and small trees, where they pursue their insect prey and obtain their water by licking raindrops or dew from the foliage. Anoles appear to adjust readily to the presence of human beings and are as likely to be found among landscape plantings around houses, even in cities, as in more undisturbed areas.

Male anoles are territorial, which means they defend a certain area of their habitat from intrusion by other males of the species. However, the males rarely fight. Rather, an anole signals "ownership" of a certain area by extending its pink throat into a fan-shaped structure, then slowly raising and lowering its head. This behavior usually results in the intruder's retreat to another area without further interaction between the two males.

The throat fan is also used as a signal between males and females as a part of the mating ritual. After mating, the female anole deposits two to four eggs in leaf litter, under pieces of bark, or in cracks and crevices. The eggs hatch in a few weeks, and the young take up an existence among the vegetation similar to that of the adults.

Reprinted from *Science and Children*, David C. Kramer, March 1985, p. 22.

John Coulter

Suggested Observations, Activities, and Questions

- Observe and describe how anoles catch and eat their food.
- Observe and describe how anoles move their eyes.
- Place an anole on different colors of construction paper put into the bottom of a jar. Does the anole change color?

- Does the anole change color in a bright environment. What about in a dark one?
- Place two male anoles in the same terrarium. How do they behave when they encounter each other? How do the anoles resolve their territoriality?

Housing and Care

For most animals in this series, I have recommended a fairly sterile environment, free of plants and other obstructions that make cleaning difficult and that can be easily overturned or uprooted. Anoles, though, seem to prefer and probably benefit from an attractively planted terrarium with plenty of twigs on which to climb. Also, since anoles are not native to most parts of the United States, it probably will not be possible to keep specimens for a few weeks and then release them into their natural habitat; thus, anoles require more permanent quarters that simulate their natural environment.

A 30-liter aquarium (or larger) makes an ideal enclosure for an anole or two, but it should be provided with a tight-fitting screen cover to prevent escapes and allow ventilation. Two to 5 cm of soil in the bottom of the aquarium will provide a medium in which any houseplant that is appropriate in size—except for cacti—can be planted to give the anoles a perch and hiding area. Coleus, philodendron, or wandering Jew (*Tradescantia*) are suitable as they are easily

transplanted, do well in a terrarium, and are strong enough to support anoles. A sturdy, branching twig should also be supplied for perching, basking, and displaying. Finally, while not necessary, a piece of bark, a rock or two, and some moss will add to the attractiveness of the anoles' quarters.

Once established, the anoles and their terrarium will be easy to maintain: sprinkling the plants each day with enough water to sustain them will also provide sufficient water for the anoles. Of course, the terrarium should be kept where there is enough light for plant growth.

An adequate, though less elaborate, environment can be created by putting a potted plant or two in a covered aquarium. This is also an ideal way to create temporary quarters. Again, the plants should be watered by sprinkling the foliage to provide water for the anoles.

As mentioned earlier, anoles are insect eaters, and they will accept only live, moving insects of an appropriate size. Mealworms are an excellent food and are easy to obtain, but because their exoskeleton is indigestible, it is unwise

to use them as a sustained diet. However, the white mealworm larvae, which have just shed their skins, are ideal and can be used whenever they are available. Anoles will also accept flies, small crickets, and other soft-bodied insects including small caterpillars and other insect larvae that climb on twigs and leaves. Food can simply be placed in the terrarium, where the anoles will find it, but most specimens will learn to accept wiggling insects from forceps.

One should not have to clean the terrarium often. The small, dry fecal pellets will not be obtrusive and will decompose in the soil. Of course, any dead insects should·be removed as they will not be consumed and will decay slowly.

Anoles do not hibernate but merely become lethargic during cool periods. They are reasonably tolerant of changes in temperature and can, therefore, be kept in classrooms with fluctuating weekend temperatures. Finally, although daily feeding is recommended, anoles can go without food for a day or two. A few extra live insects left in the terrarium will provide weekend forage for them.

Resources

Conant, Roger. *A Field Guide to Reptiles and Amphibians of Eastern and Central North America.* 2nd ed. Boston: Houghton Mifflin, 1975.

Mattison, Christopher. *The Care of Reptiles and Amphibians in Captivity.* Dorset, U.K.: Blandford Press, 1982.

Vessel, M. F., and E. J. Harrington. *Common Native Animals: Finding, Identifying, Keeping, Studying.* Scranton, Pa.: Chandler, 1971.

Reprinted from *Science and Children*, March 1988, p. 36

Snakes as Pets

By Hobart M. Smith

The ABC's of snake care are provision of (1) clean, escape-proof, ventilated quarters, (2) sufficient warmth, (3) an ample supply of clean water, (4) a balanced diet, and (5) prompt treatment of illnesses, deficiencies, and infestations.

Snake cages need not be elaborate. Cigar boxes can be adapted for keeping very small snakes. . . . A large square hole should be cut in the top and covered with plastic screen. A single strip of adhesive tape will fasten the lid.

Larger snakes (1–2½ feet) can be kept in 10-gallon glass aquaria having a wire lid, or in a box. If the depth of a cage (front to rear) is about three-fifths of its width, the cage will accommodate a snake about twice the width of the cage. . . . It is easiest to start with a well-constructed wooden box. Especially good are ammunition boxes. With the lid off, a number of holes should be drilled through the sides and the rear wall. . . . If the holes are so large that the snake can crawl through them they should be covered . . . with screen. . . .

A suitable snake cage or box should provide (1) as much circulation of air as possible, (2) visibility, and (3) protection from rough surfaces. Use a minimum of wire-covering, since snakes often rub their noses raw on the rough surfaces. . . . Climbing species may like a few branches braced in the box, but they are not necessary. Wood shavings, sphagnum, Spanish moss, dry leaves, sand, dirt, or newspapers are often put in the bottom of the box to facilitate removal of the feces. . . . Care should be taken to prevent ingestion of rough indigestible objects along with the food: wood shavings, for example, can kill the snake by splintering

Jenn Berg

and working through the wall of the stomach or intestine. If rough materials are used as a floor covering, the snake should be fed in another, bare-floored cage.

A receptacle into which the snake may retire out of sight is welcome for virtually all species, although it is not necessary. . . .

Sunshine is not essential for the welfare of snakes, provided theirs is a well-balanced diet. In fact, snakes are easily killed by overexposure to the sun. . . . You can provide warmth more safely by artificial light of a wattage between 75° and 85° F. . . .

During winter, the snakes in most parts of this country hibernate in the ground. In captivity, even when inside temperatures are comfortable [for humans], snakes usually go into a hibernating routine, becoming inactive and refusing food. Snakes in good condition can go without food for weeks or months without harm . . . .

Preferably keep your snake either at the proper activity temperature . . . and feed it regularly, or at the hibernating temperature [40°–55° F], and do not feed it at all.

Water must be provided at all times, as it is needed not only for the snakes to drink but also for bathing. Most snakes need to soak their bodies in water occasionally, especially before shedding. . . . Snakes drink like horses, by sucking water between the lips, held below the water. . . .

Food constitutes the most important obstacle to keeping snakes. Most species require live food. Training them to take more easily obtained food is usually a tedious chore. . . . Nonvenomous snakes used as pets fall into two categories according to their manner of subduing prey: constrictors and nonconstrictors. . . . The constrictors live chiefly on warm-blooded animals which they kill not by crushing . . . but simply by retarding the rhythmic movements indicative of life—breathing and heartbeat. Shortly after movements cease, the snake uncoils, carefully checks its prey to be sure it is dead, then proceeds to swallow it. . . . The nonconstrictors simply catch and swallow their prey, in some cases using their body to hold it down. . . . As a rule, snakes should have food once every week. . . .

A young snake in particular needs extra large quantities of calcium and phosphorus for bone-building. Since many snake infections center in the skin, it is essential to keep the skin vitamin (C) up to par. Snakes that eat other vertebrates, for example birds, mice, other snakes, lizards, or whole fish, get a diet balanced in all respects with these foods. Earthworms also seemingly provide a balanced diet. . . . When vitamin and mineral intake cannot be kept high enough in foods taken by the snake, they should be provided by synthetic or prepared form, mixed with the food as a powder or liquid. Cod liver oil is an excellent supplement but may not provide all needed components of the diet. . . .

Excerpted from Snakes as Pets, *4th ed., by Dr. Hobart M. Smith. NJ: T.F.H. Publications, Inc. Ltd., 1980.*

Reprinted from *Science and Children*, David C. Kramer, October 1984, pp. 34-35.

John C. Coulter

Garter Snakes

There are several kinds of garter snakes. Collectively, they are the most widespread and the best known of North American reptiles. They are found throughout most of the United States, and some range into the southern part of Canada. Although color and markings can vary considerably, a "typical" garter snake has either greenish or reddish-brown to black ground color and three longitudinal stripes—one on the midline of the back and one on each side of the body. The belly coloration is normally similar to that of the stripes. And in most species, there are two small, but distinct, light spots on the top of the head. These snakes are relatively small with most adults being from 45 to 65 centimeters (cm) long although occasionally specimens are larger.

In general, garter snakes are not unduly particular about their habitat. They can be found in wet lowlands, woodlands, meadows, relatively dry grasslands, mountain valleys, rocky outcrops, and practically any combination of habitats within their geographic range except high, barren mountains. Their diet, though, is relatively constant among the various species, with earthworms, amphibians, and fish being the mainstay. These snakes can tolerate a wide range of temperatures and are therefore active from early spring until late autumn.

Their hibernation sites include cracks and crevices in rocks or foundations, rock piles, ant mounds, and abandoned buildings. Garter snakes mate soon after emerging from hibernation and give birth to live young in late summer.

Easily captured and easily maintained as captives, garter snakes make good classroom animals. Furthermore, their presence in the room can help alleviate many uncertainties, misunderstandings, and superstitions children often have about snakes.

Housing and Care

Various kinds of cages are suitable for garter snakes, but whatever type is selected should have good ventilation, smooth interior walls, a secure cover, and adequate space; and it should be designed so that students will be able to see the occupant. Either a 10–gallon (40–liter) aquarium or an equally large wooden cage with a glass front is ideal if fitted with a secure, ventilated cover. A wood-framed screen is fine for the cover, but never use wire screen, even for ventilation holes, at floor level, as the snake will rub its nose against the rough surface and injure itself. The cage should be equipped with a 5 to 7.5 cm-deep water dish in which the snake can soak, a forked branch for climbing on, one or two rocks that the snake can rub against when shedding its skin, a small cardboard box with a 2.5-cm hole in which the snake can hide, and several

layers of newspaper on the floor.

Newspapers spread flat on the cage floor make the most suitable substrate. They are absorbent, which helps keep the cage dry (a necessity for garter snakes), and can be easily changed as needed. Sand or soil is less desirable as either of these materials makes cleaning the cage more difficult. Wood shavings should also be avoided since chips can easily be ingested along with food and cause internal injury to the snake.

Provide fresh water daily in an open shallow dish that is not easily tipped. (A 10-cm bowl is ideal.) This will provide drinking water and will allow the snake to take an occasional soak.

Diet

A captive garter snake will still prefer its natural diet, which for most species includes amphibians, fish, and earthworms. It is important, however, to exercise good judgment about the size of the food offered. All snakes swallow their food whole, and although snakes can ingest surprisingly large objects, a small garter snake cannot eat a large frog. But the size of the food is not the only consideration. Garter snakes do not kill their prey before swallowing it, and the struggle between a snake and its prey can be gruesome, especially for elementary school children. For this reason, I prefer not to use amphibians for snake food in an elementary setting.

Most garter snakes can be maintained

on a diet of earthworms, which can easily be collected or obtained from a pet store or baitshop. Simply place the earthworms, one at a time, in the cage in front of the snake. Most garter snakes will also learn to catch live minnows (about 0.8 cm long) placed in their water dish. Many specimens will also accept strips of raw fish. A sustained diet of fish, however, is inadequate for garter snakes and would need to be supplemented with earthworms.

It is difficult to recommend the proper amount to feed a snake because of the possible variations in the size of the snake and the kind of food being offered. So determine amounts through experimentation, but keep in mind that snakes rarely overeat. Finally, snakes do not need to eat daily, and a good meal once or twice a week is adequate. This, by the way, makes garter snakes especially easy to keep over weekends and holidays (though you must be sure that they never run out of water).

About Shedding

All snakes shed their skin as they grow. During periods of rapid growth, this may occur as frequently as once a month. Just prior to shedding, a snake's eyes become cloudy, it typically refuses to eat, and it may become more lethargic than usual. Shedding begins around the mouth, and the snake will facilitate the process by rubbing against the rocks or forked stick. When the skin becomes caught on a rough projection, the snake will simply crawl away leaving the skin turned wrongside out. Be sure to save the skin for further examination.

When the Project Is Over

Unless there is a reason to keep it longer, the snake should be released in its natural environment after two or three weeks.

Suggested Observations, Activities, and Questions

- Why does the snake stick out its tongue?
- Does the snake ever close its eyes?
- How does the snake move?
- How does it catch and eat its prey?
- How does the snake feel?
- If the snake does shed its skin, examine it carefully. Note the individual scales including the eye scales.

A Note on Snake Handling

For many children, handling, or even just touching, a snake can be a very positive experience; but if a child is fearful, a teacher should not press the issue.

Teachers should also be aware that although garter snakes are not venomous, their bite can be painful, and an occasional student could be allergic to their saliva.

Looking at snake handling from the snake's point of view, it should be noted that a recent study of snakes that had been kept in captivity and frequently handled by human beings showed these snakes to have adrenal glands that were completely dry. The snakes had evidently been secreting adrenaline at an above normal level for some time, and researchers infer that this is a result of the snakes' having been too close to human beings for too long. As a result of this and other studies, some zoologists do not recommend that human beings handle reptiles at all. —ED.

Bettye Sue Tubertini

A Freshwater Aquarium

Having an aquarium in the classroom lets children make first-hand observations of plants and animals and environmental interrelationships. Feeding habits, predator-prey relationships, behavior patterns, life cycles, and the effect of changing environmental conditions on populations are some activities for study.

Selecting a Container

First, determine the function of your aquarium. Do you want to keep a few aquatic organisms for a short time for a demonstration, or for individual students or for team projects? If so, very large jars, plastic containers, or battery jars are satisfactory. For more organisms that will be observed over a longer period, you will need to build a well-designed habitat of 57 to 95 liters capacity. The shape of the tank should provide a maximum of water-to-air surface area to insure an adequate supply of oxygen; i.e., don't use round fishbowls with small openings.

If the aquarium is only a holding tank for short periods, sand or plants are not necessary. Only a satisfactory water source and a compatible temperature range are needed. Most aquatic animals survive well in diffused light to darkness; avoid southern and western exposures.

Second, determine the need for mobility of the aquarium and choose accordingly. If aquaria will be moved, use only small molded glass tanks or vessels. *Never* move glass or plastic tanks with reinforced edges. Moving a tank filled with water—or allowing one to dry for a long time—distorts the frame, loosens the glass, and causes leaks.

Setting Up the Aquarium

Wash the tank thoroughly with detergent. If the tank has been used before, add a little household ammonia to the water to help dissolve grime. Rinse with clear water at least three times.

Thoroughly wash enough coarse sand or small gravel with clean water to cover the bottom of the tank to a 3 to 4 cm depth. Run warm water into a plastic bucket containing the sand. Allow the sand to settle and pour the water off until the water remains clear.

Add a 3-cm layer of clean sand or gravel to the aquarium. Add one or two clean clam shells to help neutralize the water's acidity and to provide a source of calcium for animals with shells. Add a strip of copper (approximately 2×5 cm). Copper ions released from the strip will retard the growth of single-celled and colonial forms of algae.

Add the remaining sand or gravel. Landscape the aquarium so the sand level is higher in the rear of the aquarium. This creates a trough in front for removal of dead organic matter (detritus) that will accumulate.

Add to the tank either well water, rain water, or water from springs, ponds, or streams. If tap water is used, let it "age" in uncovered containers for

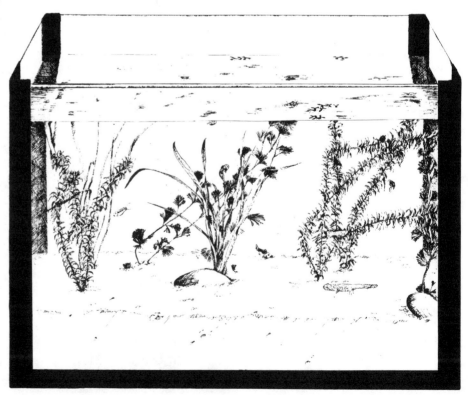

Reprinted from *Science and Children*, Carol D. and Carolyn H. Hampton, February 1979, pp. 32-34.

Aquarium Organisms

Plants

Organism	Special Uses	Care
Most aquarium plants, including those listed below:	Remove CO_2; cycle nitrogenous compounds; cover for shy or young animals; aesthetic purposes.	Space 8-10 cm apart. Provide enough light: window with northern or eastern exposure, or under fluorescent ceiling light. Temperature range: 18-24 °C. Remove excess plants to prevent overcrowding.
Elodea (*Anacharis*)	Study plant cells and chloroplasts.	Can be left floating; more attractive rooted. Push cut ends into substrate.
Fanwort (*Cabomba*)		Same as above.
Eelgrass (*Vallisneria*)		Spread roots out; press into substrate; cover with sand or gravel up to crown.
Duckweed (*Lemna*)	Study plant cells and chloroplasts; population growth studies.	Floating plants; rapid growth. Remove excess plants.

Animals

Uses and Notes

Aquatic Insects—backswimmers, damselfly nymphs, dragonfly nymphs, water striders, whirligig beetles.	Study life cycles; predator-prey relationships; niches.	Many live in water in immature stages; keep carnivores in screen cages partially suspended in water. Feed mealworms, fruitflies, mosquito larvae, *Daphnia*, other microcrustaceans, raw lean beef suspended on a string, small tadpoles, fish.
Waterfleas (*Daphnia*)	Food for small fish, tadpoles, hydra; use to clear green aquarium water (algae); study effects of environmental factors.	Feed on bacteria; nonfilamentous algae; mashed hard boiled egg yolk. Optimum temperature: 24-26 °C.
Fish—Freshwater: minnows, darters, sunfish, mosquito fish.	Study behavior: courtship, reproduction, parental care, aggression, niches.	Feed powdered fish food; earthworms, mealworms, brine shrimp, *Daphnia*, fruit fly larvae, *Tubifex* worms. Optimum temperature: 18-21 °C.
Goldfish	Study behavior, genetics, ecology. *Not recommended for general classroom aquarium:* feed on plants, stir up bottom substrate, cause heavy detritus buildup, algae growth.	Feed commercial fish food; same as "Fish." Optimum temperature: 10-25 °C; 16-21 °C for breeding.
Guppies	See "Fish".	Feed commercial fish food, brine shrimp. Adults eat young—remove fry to separate container or place females in brood chamber before bearing young. Optimum temperature: 23-27 °C.
Freshwater mussels	Study anatomy and physiology, niche, role in food web. Relatively inactive. Long-lived. Filter feeders.	Feed on phytoplankton (algae), Zooplankton (protozoans and microscopic invertebrates), organic wastes. Dead if shell remains open more than 1 mm.
Salamanders (aquatic species adapt well to aquaria):	Study behavior, niche, anatomy, and physiology, reproduction, development, regeneration of limbs.	Feed white worms, snails, chopped earthworms, small crayfish, chopped meat or liver. Sometimes eat own young.
Ambystoma (Axolotl larva) and *Amphiuma*	Adapt well to aquaria.	Optimum temperature: 21-25 °C.
Necturus and *Notophthalmus* (newt)	Adapt well to aquaria.	Optimum temperature: 18-20 °C.
Snails (including species listed):	Scavengers. Study egg development, anatomy, physiology; food for fish, crayfish, salamanders. Snails other than species listed, not recommended for general classroom aquaria: may feed on aquarium plants or produce excessive waste.	Keep in aquarium with sandy bottom and plants. Provide calcium carbonate for shell growth (crushed limestone, powdered cuttlefish bone, piece of hardened plaster of paris). Feed on encrusting algae, aquatic plants, lettuce leaves, dried fish food.
Pond snails (*Lymnae* and *Physa*)		
Mystery snail (*Planorbis*)	Changes color under changing environmental conditions; used for crossbreeding experiments.	
Tadpoles	Keep down algae growth; study life cycles, development, growth, role in food chain.	Feed on algae, aquatic plants, dried fish food, chopped lean meat.

at least three days before using so gaseous chlorine escapes.

Place a saucer or square of plastic on the sand. Pour the water in *slowly* so the sand is not disturbed. When the water is 15 to 20 cm deep, add plants.

Green Plants

Green plants make an aquarium more attractive. They provide cover for shy or young fish, remove some of the nitrogenous products released by animals, and absorb carbon dioxide.

The role played by plants in oxygenating aquarium water has been exaggerated. If there is enough air/water surface area, the oxygen concentration in the water will eventually reach the saturation point. Since the aquarium plants also use oxygen, they may actually compete with animals for what is available.

Plant the longer plants near the rear and the shorter sprigs near the center of the aquarium. Leave enough space in front for examining and handling animals and for removing detritus. Allow 8 to 10 cm between any two plants. The hardiest and most commonly used plants for classroom aquaria are listed in the Table. Collect these species from their natural habitats or purchase them from pet stores or biological supply companies.*

Fill the aquarium with water to within 2 cm from the top. Add several rocks (smooth sandstone or granite) for a scenic effect and cover for shy animals. Add a glass cover to reduce evaporation and contamination by dust and unwanted microorganisms. Glue small pieces of cork to the rim at each corner to leave space for ventilation.

Animals

The aquarium can be stocked with animals after several days. Give the water time to clear, reach room temperature, and dissolve adequate oxygen.

In selecting animals, remember: (1) predators should be isolated from prey species until you want to study feeding behaviors; (2) animals that normally live in still ponds or quiet streams will adapt better than those from running streams because of the oxygen supply; (3) avoid animals that stir up the bottom sand or uproot vegetation; and (4) using 4 L. of water per 2.5 cm animal

---
* See Resources.

prevents overcrowded conditions.

Native freshwater animals can survive in a classroom aquarium. Animal selection may vary with the students' ages, the science curriculum, organisms collected during field trips, the time of the year, the area in which you live, or your own preferences. The Table lists specific organisms with directions for their care. Mature or older specimens are generally less hardy than young animals. Fish that are 3 to 5 cm in length are best to use. Several small animals are preferred to one large animal. A suggested grouping for a 38-L unaerated, unfiltered aquarium:

15-20 rooted and/or submerged plants

2 × 2 cm of duckweed

10 small snails (or 5 large snails)

8, 2.5 cm fish (or 4, 5-cm fish)

4 tadpoles

1 newt or larval salamander

1 mussel

Put one snail in the aquarium for each 4 L of water. Scavenger snails help remove sedimentary wastes and algae. Too many snails add excrement that is unsightly and supports algae growth. Avoid species that eat rooted plants. (See Table, page 33.)

When adding fish or other animals to a tank, allow them to remain in a plastic bag containing the water in which they arrived until the water temperature in the bag is the same as the tank's. Tip the bag to empty organisms into the tank.

Maintenance

Keep an aquarium with plants in medium light. Strong light favors algae growth. A window with a northern or eastern exposure (or a fluorescent ceiling light) provides the best light.

Optimum temperature range will vary with organisms, but room temperature is usually satisfactory. (See Table.)

A common problem with classroom aquaria is overfeeding. Overfeeding leads to buildup of organic wastes, cloudy water, lowered oxygen concentration, and in some cases death. It is better to underfeed than overfeed. Remove any food that remains 30 minutes after feeding.

Daily care and weekly cleaning can slow down the processes of algae growth and accumulated organic matter. Check daily for dead plants and animals, and the accumulation of ex-

cess food. Remove the sick or dead animals and any dead foliage. Fish gulping at the surface, bubbles accumulating at the edge of the water, and dead animals are all signs of pollution or the lack of adequate oxygen.

Remove detritus that accumulates in the trough at the front of the aquarium with a siphon or a baster. By hand, remove filamentous algae each week.

Many aquarium aids on the market enhance the use of aquaria for specific objectives and organisms. Filters, aerators, heaters, thermostats, and artificial lights are your option. Accessories increase the cost of keeping a classroom aquarium. We have had success in maintaining aquaria without them.

Resources

Aquatic Insects in the Laboratory. Turtox Service Leaflet 14 (New Series). Macmillan Science, Chicago, Illinois. 1974.

Behringer, Marjorie P. *Techniques and Materials in Biology*. McGraw-Hill, Inc., New York City. 1973.

De Filippo, Shirley. "Aquarium Problems: How to Solve Them." *The Science Teacher* 42:56-57; May 1975.

Hampton, Carolyn H., and Carol D. Hampton. "Elodea." *Science and Children* 16:27-28; November/December 1978.

Hampton, Carolyn H., and Carol D. Hampton. "Snails." *Science and Children* 16:42-43; January 1979.

A Sourcebook of Information and Ideas. Nuffield Junior Science, William Collins Sons & Co., Ltd., London, England. 1967.

How to Feed Animals in Aquaria and Terraria. Turtox Service Leaflet 12 (New Series). Macmillan Science, Chicago, Illinois. 1974.

Orlans, Barbara. *Animal Care: From Protozoa to Small Mammals*. Addison Wesley Publishing Co., Reading, Massachusetts. 1977.

Plants For the Fresh-Water Aquarium. Turtox Service Leaflet 11 (Original Series). Macmillan Science, Chicago, Illinois. 1960.

Starting and Maintaining a Fresh-Water Aquarium. Turtox Service Leaflet 5 (Original Series). Macmillan Science, Chicago, Illinois. 1960.

Swenson, William, Charles Barman, and Rudy Koch. "Aquaria on a Classroom Budget." *The Science Teacher* 43:35-36; March 1976.

A Saltwater Aquarium

B ring the sea to your classroom by setting up a saltwater aquarium. While a marine aquarium is not as easily maintained as a freshwater aquarium, it is not a difficult task. If students follow instructions carefully, they will develop new skills and experiences and will soon be observing marine organisms in their own miniature ecosystem. In a class where a marine aquarium is maintained, student interests progress from casual observation, to studying feeding and behavior patterns, to taking responsibility for maintaining the tank and organisms.

In the ocean, water composition is quite stable because of the vastness and ability of seawater to dissolve and dilute water substances. Water in any given area is constantly circulated by tides, currents, and wave action. The key to a successful aquarium is maintaining good water quality.

Selecting a Tank

Use an inert tank of all glass or Plexiglass construction. A recessed ridge on which a glass or Plexiglass lid can rest will allow splashed water and condensation to drip back into the tank. Tanks of 75 to 110 liter (L) capacity are recommended. Smaller tanks do not allow for a variety and a good balance of organisms. Two 75 L tanks can hold more animals and provide more ecological niches than one 150 L tank. If one tank becomes contaminated, a second one can function while the first is cleaned.

Filtration System

Marine organisms (especially fish and motile invertebrates) produce a lot of waste material consisting of ammonia, urea, and carbon dioxide (CO_2). Ammonia is of greatest concern because it is toxic in relatively low concentrations.

There are two basic types of filtration: biological and mechanical. The essential filter for a marine aquarium is the undergravel filter, also called a biological filter because it relies on bacteria. As soon as the tank is set up, nitrifying bacteria and other microorganisms attach to the gravel and filter substrate surfaces. The microbes extract waste products and decaying matter from the water as it filters through the gravel. Ammonia and urea are converted to less toxic compounds called nitrites and finally to end products called nitrates. Nitrates, while still toxic, are only lethal at high concentrations. Green algae take up the nitrates and convert them into plant biomass. Nitrates also may be removed by a mechanical filter. (See Figures 1 and 2.)

A mechanical filter is not an absolute necessity. In conjunction with a subgravel filter, it will improve the quality of the water by removing particulate matter and keeping the dissolved oxygen concentration high. Usually this type of filter is a box attached to the outside of the tank. The water moves down through a layer of polyester fiber which filters out suspended particles and then moves through a layer of activated charcoal which adsorbs organic molecules. This type of filter in time becomes a biological filter, as nitrifying bacteria attach to the fibers.

Both types of filters must be attached to an air pump to make them operate. For 75 L tanks or larger, two undergravel filters are needed. If a mechanical filter is used, one is sufficient.

Water

Water for the aquarium may be either natural seawater or a medium made from synthetic sea salts. If you are located near the coast, marine organisms you collect will survive well in the water from their collection site. Sea water near the shore has a tendency to be polluted. Collect the water from a rock jetty or wade out into deeper water at high tide. Transport the water in plastic buckets. A good practice is to collect extra water to store for making

Figure 1. Mechanical Filter

Figure 2. Biological Filter

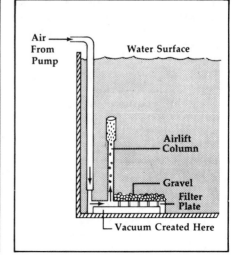

Reprinted from *Science and Children*, Carol D. and Carolyn H. Hampton, October 1980, pp. 30-32.

water changes in maintaining the aquarium. Store the buckets in a darkened place and tighten the lids to prevent algae and other unwanted contaminants from growing. The quality of the water improves with storage. Before using natural seawater, filter it through a funnel packed with polyester fiber.

You can make synthetic seawater by mixing prepackaged salts and minerals with distilled or aged tap water. Before using tap water, leave it in an open plastic bucket for three days to allow the chlorine to escape. For best results, use a well-known sea salts product from a reputable supplier. Beware of off-brand products at the pet stores.

Bottom Material

Good materials to use as a substrate are calcareous gravel, limestone pebbles, coquina shells, or crushed coral. Slow dissolution of the calcium carbonate in shells and other limestone materials will help stabilize the pH of the water. Never use silaceous gravel, sand, or colored pebbles sold in pet stores. Be sure the particle size is greater than the holes in the subgravel filter.

Run freshwater over the bottom material in a plastic dishpan or bucket to clean it. Stir the material so dirt and organic debris float to the surface. As the heavy particles settle, pour off the floating material. Continue this procedure until the water remains clear when the gravel or shells sink to the bottom.

Setting up the Aquarium

Place the aquarium tank on a sturdy level surface. Select a location out of direct sunlight to prevent excessive algal growth. Never move a tank once it has any substrate and water in it or you will loosen the seams.

Install the subgravel filter connecting air hoses and an air pump according to the manufacturer's directions. Cover the filter with 5 to 7 centimeters (cm) of substrate material.

Decorate the seascape with pieces of coral, dried sea shells, and seafans. If you use rocks, be sure they have no holes, or have holes only large enough to permit recovery of animals that might hide in them and die. Soak seascaping materials in seawater for 48 hours before placing them in the aquarium. Do not use metal objects or metal containing rocks because they may add toxic metallic ions to the water.

Place a shallow saucer on the substrate and fill the tank 2-3 cm from the top. Pour water onto the saucer to prevent gouging out the substrate material. With a wax crayon or piece of masking tape, mark the water level on the outside of the tank. Install the outside mechanical filter, if one is used. Turn on the air pump and adjust the valves. The bubbles should rise rapidly but separately through the tubes of the subgravel filters.

System Stabilization

Let the aquarium system operate for at least three days before adding any organisms. Add one hardy organism for a one-week period, then add others. This procedure will help build up nitrifying bacteria in the undergravel filter and stabilize the physio-chemical environment. If it is not possible to condition the tank with a single organism, then let the aquarium stabilize for two weeks before adding a full population.

Stocking the Aquarium

In a classroom aquarium where temperature is difficult to regulate, the best specimens to maintain are those from the Middle Atlantic and the Gulf states

Figure 3.

Feeding Schedule		
Recommended Organisms	**Food**	**Schedule**
Carnivorous Invertebrates Crabs, horseshoe crabs, starfish, brittle stars, mantis shrimp, moon snails	Feed chopped fish, small minnows, shrimp, earthworms, white worms, bits of lean beef placed at or under the animal.	Once/week
Sea anemones, coral polyps	Crush food with a little sea water into a watery paste. Pipette over the tentacles and polyps.	Three times/week
Filter-Feeders Barnacles, scallops, clams, hydroids, sponges, marine worms, sea squirts	Feed finely powdered fish food, *Daphnia*, freshly hatched brine shrimp, algal flagellates.	Three times/week
Herbivorous Species Sea hares, sea urchins, turbo snails	Feed natural growth of algae in tank, small pieces of lettuce or spinach.	Once/week
Carnivorous Fish Small fish such as killifish, sculpins, sea robins	Feed dry fish food. Feed living brine shrimp or *Daphnia*.	Daily Two times/week
Scavengers Fiddler crabs, hermit crabs, cleaning shrimp, sand dollars, mud snails, flounder, pipefish	These species will aid in cleaning the aquarium of dead and decaying materials.	

Salt Water Aquarium Maintenance Checklist*		
Daily	**Weekly**	**Monthly**
___ Adjust pump to run smoothly	___ Check water level	___ Replace 1/4 volume of sea water
___ Unclog air lines	___ Check salinity	___ Rinse polyester fibers
___ Assure airstones deliver tiny, even bubbles	___ Check pH	___ Change activated charcoal
___ Remove any dead organisms	___ Clean algae off inside walls	___ Oil filter pump motor
___ Remove any excess food	___ Gently stir up bottom debris	
___ Check water temperature	___ Change animals' diet	
___ Wipe salt off top and sides		
___ Check list of organisms to see if any are missing		

Figure 4.

(North Carolina to Texas). These specimens tolerate a temperature range of 18° to 21° C. Most biological supply houses advertise a composite of marine organisms for classroom aquaria. If you live near the coast, collect your own specimens. Thoroughly clean and rinse two milk cartons. Cut off the tops and arrange the cartons in a large ice chest with seaweed in the bottom of each carton. Fill the cartons with seawater to just below the surface of the seaweed. Add one to three invertebrate specimens to each carton. Be careful not to submerge them in water. The most common reasons for specimens not surviving transportation are overcrowding and lack of oxygen. The wet seaweed will keep specimen's respiratory surfaces moist. If the animals are not completely covered with water, more oxygen will spread across their respiratory surfaces. Place fish in a carton with just enough water to cover them. Collect specimens that are less than 5 cm in diameter. Do not pack predators with prey nor active organisms such as crabs with delicate animals. The biomass carrying capacity of a tank depends upon a number of factors including the efficiency of the filtering system and the metabolic rates of individual organisms. A good rule of thumb is about 18 to 20 animals per 75 L tank.

Acclimatizing the Specimens

Float the bags or cartons holding the animals in the aquarium. Add a little water from the aquarium to the bags every 15 minutes. If the animals act nervous or contract, stop and wait before adding water more slowly. When the water temperature in the containers equals the aquarium's, remove the specimens with a dip net or with gloves and place them in the aquarium.

Feeding Rules

1. Cut off the power supply to the filters so small food particles will not filter out of the aquarium.
2. Do not give more food than can be eaten in one to two hours.
3. Keep a few scavenger animals to help clean up excess food. (See Figure 3.)

Maintenance

Light. Do not expose the aquarium to direct sunlight. Excess light will cause undesirable algal growth and high temperatures. Diffuse light from a north or east facing window is best. If the area is exceptionally dark, use an aquarium hood with fluorescent lights.

Temperature. Extreme body temperature changes lead to stress, burrowing, dormancy, disease, and perhaps death of the animals. Keep the aquarium at 21° to 24° C.

Salinity. The salinity (salt content) should be kept at 30 parts per million or a hydrometer reading of 1.025. Once the aquarium is working and the water level is marked, add distilled or aged tap water to keep the water level stabilized. Never add more seawater; this increases the salinity. A glass cover over the aquarium will retard the rate of evaporation.

pH. The pH should be kept between 8.0 and 8.3. You can order a simple test kit from a biological supply house for this purpose. Calcareous shells and a natural growth of algae will help stabilize the pH. If the pH becomes too acid or falls below 8.0, remove a cup of aquarium water, add one teaspoon of bicarbonate of soda, mix well, and pour it slowly into the tank. If the water is still acid the next day, repeat the procedure until the water is the proper pH.

Water Changes. Make a partial water change once a month to dilute built-up wastes, replenish trace elements, and help maintain alkalinity. Siphon off one-fourth of the tank's volume and replace it with fresh seawater. Remove the mechanical filter once a month. Rinse the polyester fibers in fresh, aged tap water and replace the activated charcoal. Be sure to rinse the new charcoal to remove dust particles that could cloud the water.

Most detritus (partially decayed organic matter) will be drawn into the gravel where bacterial action will break it down. However, with time, some debris may accumulate on the substrate. Periodically stir and circulate this material so filter feeders use it. Remove large, accumulated materials with a kitchen baster. Each week, scrape algae from the inside aquarium walls with a plastic aquarium scraper. Encourage children to keep a log of their organisms and sources of food. This will be useful information in planning the composition of organisms for future aquaria. The maintenance checklist may be posted near the aquarium with a list of the organisms. (See Figure 4.)

References

1. Bower, Carol E. *Keeping A Marine Aquarium-A Guide for Teachers.* The Children's Museum of Hartford, Hartford, Connecticut. 1975.
2. James, Daniel E. *Carolina Marine Aquaria.* Carolina Biological Supply Company, Burlington, North Carolina. 1975.
3. Straughan, Robert P. *The Salt-Water Aquarium in the Home.* 4th Rev. Ed. A. S. Barnes and Company, New York City. 1976.
4. Waters, Barbara. *Small Oceans.* 4-H Marine Education Project. University of Massachusetts, Amherst, Massachusetts. 1977.

*Adapted from Crenshaw, Neil. *Starting and Maintaining a Marine Aquarium.* Florida 4-H Marine Program, Florida. 1979.

Recycling Plastic Containers for Science Activities

A wealth of useful science teaching materials can be made from discarded plastic objects with little effort and expense. Each day, tons of plasticware end up in landfills and garbage dumps of cities and towns. Plastic remains unaffected by decomposed organisms and is not recycled through the biosphere. We have saved and used all kinds of plasticware for years in our science classes. Here are some suggestions to consider for your classroom:

Clear Plastic

Transparent plastic bags make ideal "miniature greenhouses" for starting plant cuttings and growing seeds. They may be used as temporary aquaria for transporting aquatic plants and animals from the field into the classroom. Food packaging, clothing, and freezer bags offer unlimited sources of bags.

Plastic shoe boxes may be used as small animal homes and terraria, or for small animal behavioral studies. For example, tape strips of cardboard to the inside of a plastic box to construct a maze. Clear shoe boxes are good storage containers for small items which may be lost easily in drawers and on shelves.

Clear sweater boxes make excellent seed germination chambers. Place seeds between wet paper towels and store the chamber in a dark place or in indirect light. In two to three days, study the developing seeds.

Plastic vegetable crisper boxes from refrigerators will hold small mammals, amphibians, and reptiles while you study them. Be sure to drill air vents in the cover to prevent suffocation, however.

Plastic glasses and tumblers can be used for planting seeds and observing root growth. They also make good observation chambers for small animals, such as insects, worms, and isopods, collected on field trips. If you need measuring devices, calibrate these cups and glasses in metric units and use them to supplement the classroom supply of beakers and graduated cylinders.

Use plastic sheeting to cover wooden frames, making an indoor classroom greenhouse. (See pictures.) Place the plastic covered frame on a table in front of a window. Plants grow well inside this chamber even during vacations when the heat is turned off in the building. Another use of the sheeting is lining trays to make them watertight for growing plant cuttings and seeds. Use black plastic sheeting draped over plants to create periods of darkness.

Clear carbonated drink bottles (two liter size) have a variety of uses. Remove the necks with a sharp knife or scissors and use them as funnels. Remove the top of the bottle and use the bottom part as an aquarium or terrarium. Drink bottles make useful collecting containers on field trips and display chambers for specimens in the classroom.

Translucent Plastic

Milk and juice jugs (all sizes) offer many possibilities. (See pictures.) Quart jugs may be cut to form scoops for handling soil and sand. Half-gallon, gallon, and two liter jugs may be cut so that the handles remain, making them usable for collecting and transporting specimens during field trips. Gallon jugs make handy storage containers for stock solutions, liquid fertilizer, and aged tap water for aquaria. All sizes of jugs may be used for storing noncorrosive liquids and solids. Gallon jugs with the tops removed make good temporary aquaria and terraria for small or individual specimens.

Use poultry basters to remove unsightly debris from the bottom of aquaria and to pick up and transfer small aquatic organisms to other aquaria during cleaning sessions. Basters are also ideal equip-

Reprinted from *Science and Children*, Carol D. and Carolyn H. Hampton, January 1981, pp. 42-43.

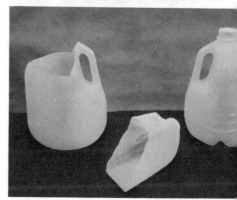

ment for capturing small aquatic animals in streams and pools while on field trips.

Opaque Plastic

Use plastic garbage cans for storing potting materials; e.g., gravel, soil, sand, and vermiculite. Use plastic scoops made from quart milk jugs to transfer materials for storage inside the cans. Garbage cans may also store large objects and pieces of equipment which need to be protected from dust and damage.

Plastic foam ice chests make good collecting containers on field trips and storage chests for equipment and supplies in the classroom. They may serve as homes for small animals unless the animals can gnaw through and escape. Eight to 10 frogs may be kept indefinitely in a large plastic ice chest. If the chest can drain into a sink, adjust the flow of water into the chest so freshwater stays at a suitable level for the frogs. Several bricks stacked at one end will allow frogs to leave the water, but remain on a moist surface. Drain the chest for feeding the frogs and periodic cleaning.

Plastic foam cups make excellent pots for growing seeds and cuttings in the classroom. Small animals may be collected and observed in them during field trips. They make good homes and hiding places for small organisms when placed in terraria and aquaria.

Plastic egg cartons make excellent chambers for starting seeds in early spring. Place potting soil and seeds in each cavity. When the seedlings are large enough, transplant them outside to a garden or flower bed.

Use sheets of plastic foam as display boards, pinning objects to them for drying and holding their shape. String art such as simulated spider webs may be attached to the sheets with pins or small nails. Cut in pieces, the sheets may be used as partitions in cages, aquaria, and terraria for separating animals.

Garbage bags may be used for collecting and transporting plant specimens on field trips. They make convenient "dust covers" for large pieces of equipment, such as microprojectors, which must be left out in the classroom.

Use plastic buckets with lids for storing aged tap water. The lids should be left ajar for two days to allow the unwanted chlorine gas to escape, then covered tightly to keep out contaminants. The buckets are also useful for carrying specimens collected on field trips. Do not place the lid on too tightly when small animals are being transported or they may suffocate.

Many articles in *Science and Children* and *The Science Teacher* suggest uses for plasticware. A few of these are listed in the references.

References

Gilmore, Virginia. "Coca-Cola Bottle Terrarium." *Science and Children* 16:47; April 1979.

Hampton, Carolyn H., and Carol D. "The Establishment of a Life Science Culture Center." *Science and Children* 15:7-11; April 1978 (p. 7 of this book).

Padilla, Michael. "Chasing Bugs." *Science and Children* 16:8-9; May 1979.

Riley, Joseph, and K.D. Sowinski. "Natural Partners: Science and Reading." *Science and Children* 17:46-47; October 1979.

Snyder, Linda L. "Short on Measuring Equipment? Make Your Own!" *Science and Children* 16:12; May 1979.

Seshadri, Ven. "Overflow Jar." *The Science Teacher* 47:44-45; May 1980.

Siddons, J.C. "Jack and Jill and Other Homemade Electrophoruses." *The Science Teacher* 46:42-44; September 1979.

Wulfson, S., S. Zucker, R. Konick, and R. Fetzer. "Sporks." *Science and Children* 15:25-27; Nov/Dec. 1977.

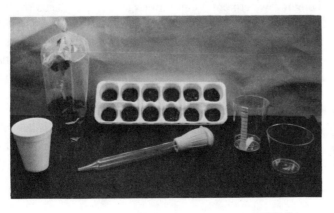

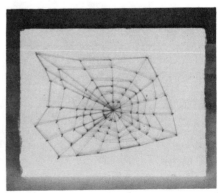

Reprinted from *Science and Children*, February 1991, p _.'-29

A Big Lesson In a Small Pond

By Allan Friedman

Fifteen bottles of muck mixed with teacher ingenuity can bring a pond to the schoolyard.

I HAD ALWAYS WANTED TO study ponds with my students but felt at a loss because any hands-on activities would require access to an authentic pond—a difficult proposition in most cities. Last year I built a box in order to raise crawfish in my classroom. This project proved quite successful, so I stored the box hoping it would be of future use. Then, while working on a botany curriculum and again feeling the need to study a real pond, I realized that the same box could be set up outside the classroom for just that purpose.

Building the Pond

To construct the pond I made a 1 x 0.6 x 0.35-m box with ½" plywood strips and reinforced the corners by nailing the plywood to 2 x 2s. Next, I bought a waterproof nylon tarp large enough to fit into the box and overlap the sides somewhat. I secured the tarp to the box with 12 small metal clothespins.

I then filled the pond with tap water and allowed it to sit for a week. My students and I added as much driveway and aquarium gravel as we could find. I located an old pump and used it to recirculate the water. (This is not necessary, but it slows the growth of algae.) Students visited local ponds and brought to class as many bottles of "pond muck" as they could scoop up from the bottom. We collected about 15 bottles and dumped these into our pond. It took another week for the muck to settle and the water to clear. A few students brought in some large rocks that we placed at one end of the pond to make a platform. We built the rocks up out of the water to create a small waterfall with the outflow from the pump.

Adding Some Life

We then looked at our pond water under a microscope and found very little life. At this point I bought two pond plants, water four-leaf clover (*Marsilea mutica*) and marsh marigold (*Caltha palustris*), from a local nursery and placed them in the pond, underwater, in their pots. I also ordered additional plants and animals—freshwater mussels, snails, tadpoles, aquatic ostracods, aquatic isopods, damselfly nymphs, dragonfly nymphs, salamander larvae, and catfish. (Check the *1991 Supplement of Science Education Suppliers* that accompanies this issue to locate sources for stocking your pond.)

(Above) Students monitor the pond daily to check for changes. (Left) Aquatic plants thrive in the warm spring weather.

Our plants began to grow, and the pond developed a greater concentration of microscopic life. As the animals arrived, over a period of weeks, we put some in the pond and temporarily kept others in glass containers so that we could spend time easily observing their behavior. We divided about 30 tadpoles among 10 jars, and pairs of children were assigned to each jar to maintain, clean, feed, and observe them. The tadpoles raised in the classroom were fed artificial food while those in the pond had to fend for themselves.

By the time we had placed all the animals in the pond, the new ecosystem seemed to be maintaining itself quite well. Plants were growing, animals were thriving, and altogether the system seemed balanced.

Experiencing the Weather

During the spring we witnessed a variety of weather conditions and the subsequent effects on the pond environment. Soon after the plants and a few of the first animals went in, our area experienced an unusual cold spell, with temperatures well below freezing for a number of days. The freeze was cold enough to burst some of our water pipes and to affect the pond. One catfish and several other animals died. The plants were severely damaged-most of the duckweed died, and the pond lettuce (a floating plant) lost most of its leaves and a great deal of its root system.

In early April, our area experienced a week of temperatures above 32°C. This produced a period of incredible growth. Our marsh marigold flowered prolifically, the four-leaf clover's roots grew out of the pot, the water lettuce added a half dozen new heads, and the duckweed covered almost all of the rest of the pond. The plants grew so much that we had to do some clearing in order to continue our observations. This variety of conditions (the temperature ranged from -3.9°C to 35°C) provided valuable lessons for all of us.

Diving into Activities

Since we built our pond, we have done pond-related activities one to three times a week. A sampling of those activities follows.

1. Small groups or pairs of students observe specific animals or plants and then report their observations to the whole class.

2. Using the microscopes fairly often, students should draw and identify what they observe. Notice that the concentration of microscopic organisms increases greatly over time.

3. Compare tadpoles living in the pond to those in the glass containers. Which grow faster, appear healthier, and develop into frogs sooner?

4. Take pond field trips. Go to a local park for a Pond Day. Drag the pond with homemade tools. Students can then identify and study what they find before releasing it.

5. Monitor the pond daily, and if anything is new or changed, the students must try to find it.

6. Read pond books and write pond stories.

A new discovery is made.

7. Draw the pond at various times. We draw the pond iced-over from top to bottom, from memory, and we map the top. Create pond animals out of clay.

8. Maintain the pond over summer vacation, and when students return, study what happened. Eventually the class will learn about the system during a full year.

Continuing Interest

This successful project was simple to build and cost only about $50 (not including plants and animals). There were no problems with vandalism aside from the occasional appearance of an apple core or sandwich wrapping. As a matter of fact, the entire student body and many parents showed a continuing interest in the pond's status. Hardly a person walked by without peering into the pond and asking about it.

My students have maintained a very high interest the whole time, and we continue to learn. They are proud of

the pond and like to show it to visitors, explaining what they have done.

We all have had a chance to carefully observe and participate in an ongoing, fairly natural system being built, set up, stocked, managed, and finally maintaining itself.

Resources

Amos, W.H. (1967). *The life of the pond.* New York: McGraw-Hill.

Back, C., and Watts, B. (1984). *Tadpole and frogs.* Morristown, NJ: Silver Burdett.

Carolina arthropods manual. (1982). Burlington, NC: Carolina Biological Supply.

Cox, G. (1988). *Pond life.* New York: Michael Kesend.

Jennings, T. (1985). *The young scientist investigates pond life.* Chicago: Children's Press.

Kuhn, D. (1988). *The hidden life of the pond.* New York: Crown.

McClintock, T. (1938). *The underwater zoo.* New York: Vanguard.

Mitchell, J., and the Massachusetts Audobon Society. (1980). *The curious naturalist.* New York: Prentice-Hall.

Nasco's invertebrates. (1980). Fort Atkinson, WI: Nasco.

Outdoor Biology Instructional Strategies. (1979). *Ponds and lakes.* Berkeley, CA: Lawrence Hall of Science.

Perkins, K., and Whitten, R. (1981). *Reptiles and amphibians: Care and culture.* Burlington, NC: Carolina Biological Supply.

Reid, G.K. (1967). *Pond life.* New York: Golden Press.

Svedberg, U. (1988). *Nicky the nature detective.* Toronto: R and S Books.

Teachers guide for pond water. (1976). New York: McGraw-Hill.

Whitten, R., and Pendergrass, W. (1980). *Carolina protozoa and invertebrates manual.* Burlington, NC: Carolina Biological Supply.

ALLAN FRIEDMAN teaches second and third grades at Berkwood-Hedge School in Berkeley, California. Photographs courtesy of the author.